The Impact of Climate Change on Our Life

Abdelnaser Omran • Odile Schwarz-Herion
Editors

The Impact of Climate Change on Our Life

The Questions of Sustainability

Editors
Abdelnaser Omran
School of Economics, Finance & Banking
Universiti Utara Malaysia
Sintok, Kedah, Malaysia

Odile Schwarz-Herion
Dres. Shwarz-Herion KG Detective Agency
Ettlingen, Germany

ISBN 978-981-13-3993-6 ISBN 978-981-10-7748-7 (eBook)
https://doi.org/10.1007/978-981-10-7748-7

Softcover re-print of the Hardcover 1st edition 2018

Printed on acid-free paper

This Springer imprint is published by the registered company Springer Nature Singapore Pte Ltd. part of Springer Nature.
The registered company address is: 152 Beach Road, #21-01/04 Gateway East, Singapore 189721, Singapore

I dedicate this book to
PROFESSOR DR. YUSNIDAH BT IBRAHIM
Deputy Vice Chancellor (Academic and International)
Universiti Utara Malaysia
Kedah State, Malaysia

Preface

This book is written to help people to understand the impact of climate change and discuss its effect on our lives. This concerns particularly the question if humans are able to solve this problem. It is written at the most appropriate time when the impact of climate change is becoming increasingly visible all across the globe, having an impact on almost every society worldwide. The book at hand is supposed to increase awareness about the fact that the consequences of climate change do not only concern the ecological and the economic pillars of sustainable development but even directly affect its social pillar by posing a threat to the very existence of humankind. Meanwhile, it can no longer be denied that our climate has changed and will continue to change due to many factors. Factors often cited as possibly contributing to climate change are, inter alia, the increasing world population and the concomitant increase in the use of environmentally damaging energies partly due to the increasing use of electricity and modern technology in developing countries which previously had a more ecologically sustainable way of life in harmony with nature before they were manipulated to emulate the Western lifestyle at any price, partly due to Western consumption patterns becoming even more unsustainable. The controversies about suitable solutions to these problems are increasing.

The major theme of this book is on climate change. Which covers different topics related to its patterns in different countries and clarifies the issues related to it and its impact on our lives. Climate change is still, arguably, one of the most critical and controversial issues the world is facing in the twenty-first century. This book is organized into 12 chapters.

Chapter 1, starting with a UN quote about the impact of climate change on our lives, deals with the impact of the climate change discussion and the related sustainable development concept on society, science, culture and politics, describing its long way from the first Club of Rome (CoR) report "Limits to Growth" via a decade-long series of Climate Change Conferences to the recent Climate Change Conferences like COP21 in Paris with the subsequent Paris Agreement and the recent COP22 in Marrakech to current challenges. The author puts a certain emphasis on the abuse of the climate change topic by various actors. This includes opportunistic corporate lobbyists exploiting the Paris Agreement for merely economic purposes as well as fearmongers among writers, film producers and scholars hindering a sober and objective discussion about climate change by stirring up irrational panic. This fearmongering along with increased time and action pressure and radical depopulation agendas involves the danger of a global dictatorship with fascist traits based on outdated concepts and ideologies rather than seeking positive, innovative and democratic solutions.

Chapter 2 starts with the topic of natural disasters, taking references to the Kelantan flood as recent representative example to be featured in this chapter. Thus, this chapter identifies and analyses the factors which caused the catastrophic flood in Kelantan state in 2014 while putting a particular focus on one of the districts of the state, named Kuala Krai District. It has been found that equally natural and human factors contributed to the catastrophic flood in Kelantan state. Moving into the major economic and health sectors and their possible contribution to controlling climate change, Chap. 4 provides an overview of the economics of climate change and health. This chapter discusses the economic costs incurred by the effects of climate change on health, through an increased burden both on individual expenditures and the national health systems. Energy and water are inextricably connected in agricultural systems. Thus, in Chap. 5, the author discusses the role of climate change on the water-energy nexus and how efforts to increase efficiency in both energy and water end-uses can increase Queensland's agricultural sector's resilience in Australia. The chapter also discusses how efficiency in energy and water end-uses can reduce the exposure of the sector to acute and chronic stressors, including high utility bills which, with climate change, are negatively impacting agricultural productivity. In contrast, Chap. 6 puts its emphasis entirely on the issue of water security and climate change issues in developing countries by introducing partnership procurement for sustainable water projects in Nigeria. The study on which Chap. 6 is based examined the current water security condition in some selected cities in Northern Nigeria, which had been the region most affected by climate change. The study outlined the partnership management approach and further highlights the potential of partnership procurement strategies for the management of sustainable urban water supply in Nigerian cities.

Looking at the impact of waste on our lives, Chap. 7 offers a new approach on waste and the issues connected to it, such as the question to which extent waste can complicate our daily actions and influence the decay of the environment. Moreover, this chapter states that renewable materials should be sorted and separated from other types of materials to avoid that they cross-contaminate each other, to increase the value of the materials and to ease the process of manufacturing. Chap. 8 identifies the factors that could sustain the municipal solid waste management and practices in Ajdabiya City in Libya.

Chapter 9 looks at fixing climate change: accounting disclosure remedies by exploring areas in which sustainability reporting could be expanded beyond corporate environmental reporting to other necessary disclosures that will curve recklessness in the course of pursuing an economic goal. In Chap. 10, the authors from Malaysia show – from the perspective of agriculture – how human activities can increase the negative impacts of climatic change by means of their lifestyle. This chapter also tries to reduce the sometimes overexaggerated fear of climatic impacts on human beings by suggesting steps and alternatives and the possibility of international cooperative efforts. In the context of the situation in Pakistan, Chap. 11 evaluated various environmental factors and their impacts on the local tourism sector in Pakistan. Three major dimensions of the overall environment – economic and financial, social development and macroclimate changes – were selected with their key indicators based on the data sets from World Development Indicator (WDI).

The last chapter (Chap. 12) discusses anthropogenic climate change and countermeasures, focusing on the chances and risks of weather modification techniques and climate engineering (CE). The emphasis is put on the numerous and often incalculable risks of weather modification and CE technology including its potential abuse for covert warfare or other hostile purposes. This possible abuse had already been identified in the early 1960s by Harry Wexler who warned of this technology's potential to deliberately causing damage or even targeted geophysical warfare, e.g. by attacking the ozone layer over a rival nation, and also discussed this technology's capability to intentionally raise the global temperature by almost 2 °C or decrease it by over 1 °C. In any case, even those weather modification and geoengineering techniques originally developed for civil purposes can be turned into powerful weapons of war, possibly posing a much bigger threat to the planet and all life on earth than inadvertent human-made greenhouse gas emissions.

A variety of different climate change-related topics covered in this book help people to view this topical issue from many different angles, sensitizing readers for the ample challenges to human life due to climate change, ranging from changes in rainfall and high flood to drought and other extreme weather patterns and weather disasters, while also discussing the abuse of this topic for covert political and military agendas. Proposing alternative solutions, strategies and best practices such as project financing initiatives to public authorities has proven to be efficient in developed and developing countries especially in Nigeria, Australia, Malaysia, Libya, Romania, Pakistan and other countries across the world in dealing with problems caused by climate change. At times when most countries, especially developing economies, are suffering from an economic meltdown, making it difficult for them to sponsor projects, our book addresses strategies on how to deal with this challenge and is thus a must-read for faculties and schools at both local and international universities across the world.

Sintok, Malaysia Abdelnaser Omran

Acknowledgements

First and foremost, I would like to express publicly my deep appreciation to my wife for her motivation and endless support throughout my career and writing this book. I also thank my two wonderful daughters (Yara and Yusra) and my son (Yazeed) for always making me smile and for understanding those weekend times when I was writing and completing this book instead of playing games or taking them out. I hope that 1 day they can read this book and understand why I spent so much time in front of my computer.

Dr. Odile Schwarz-Herion and I collaborated to find the great and professional authors that helped us write this book. In the end, I believe that the team of authors that was chosen provides the perfect blend of knowledge and skills that went into authoring this book. My sincere thanks to all the authors for devoting their time and effort towards this book. I thank each of the authors for their effort towards this book. Thanks for everything; I look forward to writing more books with you soon.

Finding a loyal friend is as hard as finding a teardrop in the ocean. I am extremely grateful to my best friend, Dr. Odile Schwarz-Herion (coeditor of this book), who constantly took keen interest in making this book a success one. Thanks dear Dr. Odile for sharing my happiness when starting this project and for encouraging me when it seemed too difficult to be completed. I would have probably given up without Dr. Odile's support and example on what to do when you really want something. You did indeed play a crucial role in making this book worthwhile.

Speaking of encouragement, I must mention Professor Dr. Elena Durica, my faithful friend from Romania, for the great effort and for giving endless support and assistance in making this book a success by inviting some scholars to contribute. Although I am missing her contribution is this book but I will be looking forward for the opportunity to have her contribution in my next projects.

I would also like to take this opportunity to convey my sincere thanks to Professor Yusnidah Ibrahim as the former dean of the School of Economics, Finance and Banking at Universiti Utara Malaysia for her unwavering support, assistance, collegiality, and guidance over the past 3 years.

I convey my deep independence and gratitude to Dr. Mei Hann Lee from Springer (Japan Branch) for her guidance, encouragement and endless support along the way to make this project a success.

I also wish to express my gratitude to all those who have directly and indirectly contributed to the accomplishment of this book and my apologies for not mentioning them personally one by one.

Sintok, Malaysia Abdelnaser Omran

Contents

Contributors

Arpah Abu Bakar School of Economics, Finance & Banking, Universiti Utara Malaysia, Sintok, Kedah, Malaysia

Mazen Ibrahim Al-Masawa School of Economics, Finance & Banking, Universiti Utara Malaysia, Sintok, Kedah, Malaysia

Amir Hussin Baharuddin School of Economics, Finance and Banking, Universiti Utara Malaysia, Sintok, Kedah State, Malaysia

Ahmad Bello Department of Accounting, School of Business, Ahmadu Bello University, Zaria, Nigeria

Alexandru Bodislav The Bucharest University of Economic Studies, București, Romania

Florina Bran The Bucharest University of Economic Studies, București, Romania

Popescu George Cristian Faculty of Business and Administration, The University of Bucharest, Bucharest, Romania

Georgina Davis Griffith School of Engineering, Griffith University, Brisbane, QLD, Australia

Abdelsalam O. Gebril Department of Natural Resources and Environment, Faculty of Arts and Science, EL-Marj Branch, University of Benghazi, EL-Marj, Libya

Hafiz Waqaz Kamran School of Economics, Finance, and Banking, Universiti Utara Malaysia, Sintok, Kedah, Malaysia

Wen Chiat Lee School of Economics, Finance and Banking, Universiti Utara Malaysia, Sintok, Kedah, Malaysia

Norlida Abdul Manab School of Economics, Finance & Banking, Universiti Utara Malaysia, Sintok, Kedah, Malaysia

Abdelnaser Omran School of Economics, Finance & Banking, Universiti Utara Malaysia, Sintok, Kedah, Malaysia

Carmen Rădulescu The Bucharest University of Economic Studies, București, Romania

Ion Andreea Raluca Faculty of Agrofood and Environmental Economics, The Bucharest University of Economic, Bucharest, Romania

Odile Schwarz-Herion Dres. Shwarz-Herion KG Detective Agency, Ettlingen, Germany

Abdullahi Nafiu Zadawa Quantity Surveying Department, School of Environmental Technology, Abubakar Tafawa Balewa University, Bauchi, Nigeria

About the Editors and Authors

Editors

Dr. Abdelnaser Omran is currently working as a visiting scholar at the Department of Risk Management and Insurance under the School of Economics, Finance and Banking at Universiti Utara Malaysia. He is also a permanent academic scholar at the College of Engineering at Bright Star University in Libya. He has written widely on environmental engineering and waste management issues in Africa, Asia and Europe and has also contributed many other publications in the field of project and construction management, sustainable development-related issues and risk management. So far, he has published more than 250 articles in international journals and proceedings papers at local and international conferences. Besides that, he has written seven books. Abdelnaser served as an international reviewer and an editorial member for quite a number of international journals, and he also served as a scientific committee member for many local and international conferences and seminars.

Dr. Odile Schwarz-Herion studied law (major criminology) at BA level, earned an MBA with specialization in industrial engineering from the Stuttgart Institute of Management and Technology (SIMT) and got her PhD degree in economic sciences with a focus on sustainable development from the University of Hohenheim in 2005. During and after her studies, Schwarz-Herion worked as a sales engineer in large international companies and later on as an SD consultant, also acting as a keynote speaker and round table chairperson at international sustainability conferences. In 2011, a ZADcertified private detective, Dr. Odile Schwarz-Herion, who also holds a comprehensive state security licence, founded Dres. Schwarz-Herion KG Detektei, a private detective agency specializing in environmental crimes and white-collar crimes. In 2015, Dr. Odile Schwarz-Herion joined the World Association of Detectives (WAD) and was elected as a designated director to the WAD board in 2016. Schwarz-Herion is author and coeditor of several books and journal articles about SD, economics, international politics and crime investigation.

Authors

Dr. Arpah Abu Bakar is a senior lecturer at the Department of Banking and Risk Management, School of Economics, Finance and Banking, Universiti Utara Malaysia. She has a BBA in actuarial science and in risk management and insurance from the University of Wisconsin-Madison, USA. She obtained her MBA specializing in insurance from the University of North Texas, USA, and a PhD from Universiti Utara Malaysia. Her main research interests include risk management, health insurance and risk financing. She has been involved in disaster risk reduction and climate change adaptation research focusing on risk financing for several years.

Mazen Ibrahim Al-Masawa is a civil engineer, holds a master's degree in project management and currently studies PhD in risk management, with experience of more than 15 years in design, implementation and management of construction projects, especially mud buildings. He has worked in the government framework and with civil society and local community organizations. The most important companies that he has worked with were the German Organisation for Technical Cooperation Foundation, the United Nations Development Programme (UNDP) and the Yemen Liquid Natural Gas Company. He received an Aga Khan Award and is a member of the working team who was honoured at the award ceremony in Shibam.

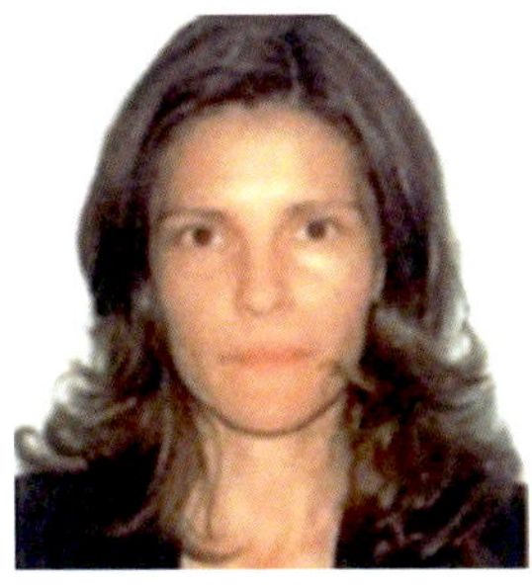

Associate Professor Dr. Ion Andreea Raluca is a lecturer at the Bucharest University of Economic Studies, Department of Agro-food and Environmental Economics. She holds a bachelor's degree in agricultural economics, a master's degree in management of agricultural and environmental systems and English communication for teaching and researching in economics and a PhD in agricultural economics. She went to the Postdoctoral Research School Knowledge Transfer Economy in Sustainable Development and Environmental Protection. Her main areas of expertise include food and nutrition security and safety, agricultural marketing and markets, agro-food chains, sustainable consumption, competitiveness, trade, etc.

Amir Hussin Baharuddin joined the School of Economics, Finance and Banking (SEFB), College of Business, University Utara Malaysia, as a visiting professor in November 2011. Before serving SEFB, he was a professor of economics at Universiti Sains Malaysia, Penang, for 36 years (1975–2011), teaching undergraduates and graduates. He had written over 100 papers and 6 books and translated 14 economics books for Dewan Bahasa dan Pustaka Malaysia. He served as the dean of School Social Sciences at Universiti Sains Malaysia from 1981 to 1989.

Dr. Ahmad Bello is an associate professor of accounting with one of the leading universities in Nigeria, Ahmadu Bello University, Zaria. His research interests include financial reporting and corporate governance. He also provides consultancy and training services in research, academic strategic planning, audit and assurance and corporate sustainability programme. Currently, he is a member of the University Board of Research and also the deputy director of the Institute of Administration.

Alexandru Bodislav is a lecturer at the Economics and Economic Policy Department of the Bucharest University of Economic Studies. He participated in numerous online and onsite courses and seminars at Harvard Business School, Columbia Business School, Michigan University, IBM and Microsoft. During his postdoctoral fellowship, Bodislav has contributed on The Big Data: Business Intelligence Synergy – The Solution for Building Relevant Indicators for an Emerging Economy. He also published and presented more than 130 research papers worldwide. Bodislav authored or co-authored several books published in the United States of America and Europe, some of them being bestsellers in their domains, especially the one published by Palgrave Macmillan.

Florina Bran is a professor at Bucharest University of Economic Studies, at the Department of Agro-Food and Environmental Economics. She is a PhD tutor for more than 12 years and acted as vice dean for the Faculty of Agro-Food and Environmental Economics. During her tenure at BUES, she developed several courses for undergraduate, graduate and doctoral level. Her skills promoted her as chair of various commissions for organizing international scientific events. She received the awards of professor emeritus, doctor honoris causa, Excellency Award and the honour of being visiting professor at Parthenope University of Naples (Italy). Professor Bran published more than 50 books and over 200 articles as author or co-author.

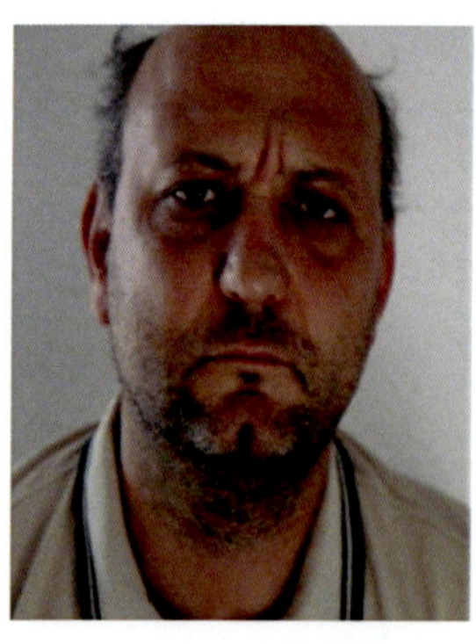

Associate Professor Dr. Popescu George Cristian is a lecturer at the University of Bucharest, Business and Public Administration Faculty; a specialist in ecological marketing, agro-food marketing, SME marketing and other marketing specialties; and an administrator or shareholder at six companies with trades in Europe, particularly the top companies in Romania in the field of organic products, and has published over 50 articles and books. He has finished several postgraduate studies on human health management and postdoctoral studies on the influence of lifestyle on human health.

Dr. Georgina Davis is a chartered waste manager and chartered environmentalist (CEnv) with 25 years' international experience. Georgina is the founder of the Waste to Opportunity Enterprise and holds an adjunct position at Griffith University. She provides policy advice to Queensland's peak farming body, the Queensland Farmers' Federation (QFF) and its member organizations; links opportunities in current government policy to operational issues impacting the agricultural sector; and provides advocacy where required. Georgina is the editor for the *Waste and Recycling Industry Association of Queensland's (WRIQ) Newsletter*, and she is a columnist of *Inside Waste*'s "Waste Opportunist" providing comprehensive information and insight once every 6 weeks around opportunities and solutions that could change and challenge the status quo. She has published over 150 trade articles, academic papers and book chapters and has presented over 60 international conference papers.

Dr. Abdelsalam O. Gebril is an assistant professor at Department of Natural Resources and Environment, Faculty of Arts and Science, University of Benghazi in Libya. His research relates to rural and urban environmental management, waste management and some geographical-related issues. He published many articles in reputable journals and proceeding conference papers in different events. In addition, he was appointed in 2014 as the representative of the Euro-Arab Chair for the Euro-Arab Association. Besides, he is currently the President of the Academy of Graduate studies in the Eastern Province located in Benghazi city in Libya.

Hafiz Waqas Kamran holds a MPhil majoring in finance and is head of finance and lecturer at the University of Central Punjab UCP Faisalabad, Pakistan. He has a teaching experience of 6.5 years. Mr. Hafiz is currently pursuing his PhD study specializing in finance (risk management) at the School of Economics, Finance and Banking, Universiti Utara Malaysia. I am done with 20 research publications and have a very good command on STATA-14, SPSS AMOS Graphics 21, R-Studio and EViews as well.

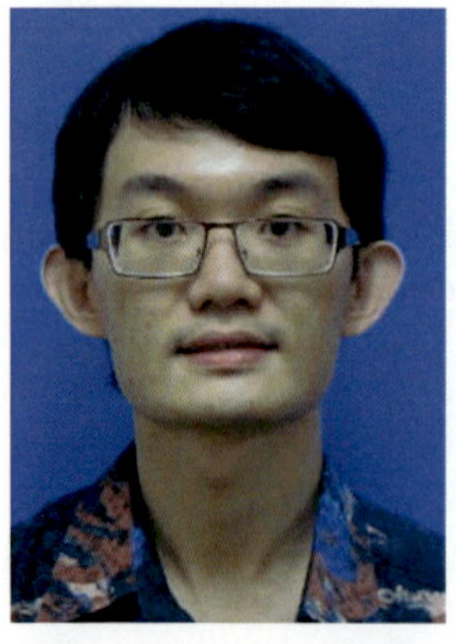

Dr. Wen Chiat Lee is currently the senior lecturer of economics at the Universiti Utara Malaysia. He received his PhD in environmental and resource economics from the Universiti Utara Malaysia in 2014. While at the UUM, he participated in a discussion on international trade, environmental economics and fisheries economics. He has completed several projects on these few areas. His college grant focuses on the environmental economics on estimating the price of carbon emissions from oil palm production in Malaysia. He has experience in international trade, environmental economics, economics on relocation, the economics of compensations and fisheries economics.

Norlida Abdul Manab is currently associate professor in the field of risk management at the Department of Banking and Risk Management, School of Economics, Finance and Banking, College of Business, Universiti Utara Malaysia. Her main research interests include enterprise risk management, financial risk management, corporate governance and other issues related to risk management. She is actively involved in teaching and supervising PhD students in the area of risk management and insurance.

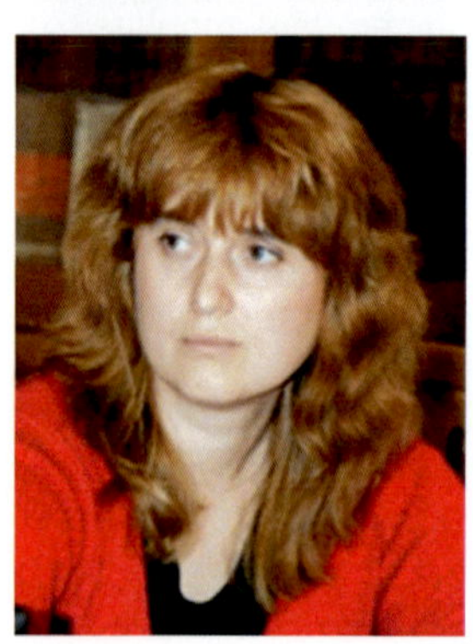

Carmen Rădulescu is an associate professor at the Bucharest University of Economic Studies, at the Department of Agro-Food and Environmental Economics. She has been a PhD tutor since 2017 for the Faculty of Agro-Food and Environmental Economics. During her tenure at BUES, she developed several courses for graduate and undergraduate levels. Her skills promoted her as chair of various commissions for organizing international scientific events and as a member of many editorial boards of international database indexed journals. She received awards for her work as a professor and also for the books she published. Professor Rădulescu published several books and more than 150 articles.

Dr. Abdullahi Nafiu Zadawa is a lecturer II at the Department of Quantity Surveying, School of Environmental Technology, Abubakar Tafawa Balewa University, Bauchi, Nigeria. He obtained a BTech degree in quantity surveying from ATBU, Bauchi, Nigeria. He also obtained master's and PhD degrees in project management from Universiti Sains Malaysia. He is a member of the Nigerian Institute of Quantity Surveyors and a graduate member of the Nigerian Institute of Management. His area of research interest includes procurement management, project management and also environmental management.

Chapter 1
The Impact of the Climate Change Discussion on Society, Science, Culture, and Politics: From *The Limits to Growth* via the Paris Agreement to a Binding Global Policy?

Odile Schwarz-Herion

Abstract The origin of the *Climate Change-by-CO_2-hypothesis* can be traced back to a study cited in the first *Club of Rome (COR)* report *The Limits to Growth* from 1972. The potential long-term impact of this report and subsequent reports to the Club of Rome (COR) in the 1970s marked the beginning of a series of Climate Change Conferences – from the *First World Climate Conference* in Geneva back in 1979 via the *UN Conference on Environment and Development (UNCED)* in *Rio de Janeiro* and the *Rio+ conferences* up to the *21st Conference of the Parties (COP21)* in Paris, followed by the *Paris Agreement* and the recent *COP22* in Marrakech. Since the Millennium, the Climate Change discussion, especially the predictions of the *Intergovernmental Panel on Climate Change (IPCC)*, has sparked controversies among scientists and scholars of various disciplines as shown, inter alia, by the so-called "ClimateGate"-scandal. Warlike Climate Change scenarios in weather disaster movies like *The Day After Tomorrow*, *Hell*, and *Snowpiercer* suggest that humans should act before it is too late, having a dramatic impact on the collective feeling that humankind is steering toward a climate catastrophe and the world is about to collapse. This fear might be exploited by those who strive for a binding global policy and the establishment of a global authority.

Keywords Climate Change discussion · UN Sustainable Development Goals · Club of Rome (COR) · Intergovernmental Panel on Climate Change (IPCC) · Conference of the Parties (COP) · Paris Agreement · Weather disaster movies · Global policy · Global authority

---

O. Schwarz-Herion (✉)
Dres. Shwarz-Herion KG Detective Agency, Ettlingen, Germany
e-mail: EOKK.Black@t-online.de

A. Omran, O. Schwarz-Herion (eds.), *The Impact of Climate Change on Our Life*, https://doi.org/10.1007/978-981-10-7748-7_1

1.1 Introduction

> Climate change is now affecting every country on every continent. It is disrupting national economies and affecting lives, costing people, communities and countries dearly today and even more tomorrow. People are experiencing the significant impacts of climate change, which include changing weather patterns, rising sea level, and more extreme weather events. The greenhouse gas emissions from human activities are driving climate change and continue to rise. They are now at their highest levels in history. Without action, the world's average surface temperature is projected to rise over the 21st century and is likely to surpass 3 degrees Celsius this century… (UN 2017a).

In this statement, the UN stresses the dramatic impact of Climate Change on all pillars of Sustainable Development (economic, social, and ecological) before leading over to the *21st Conference of the Parties (COP21)* in Paris from December 2015 and the subsequent *Paris Agreement* which entered into force on November 4, 2016 (UN 2017a; UNFCCC 2017a).[1]

The mandate to "take urgent action to combat climate change and its impacts" is listed as Goal 13 of the UN *Sustainable Development Goals* (UN 2017b). The inclusion of Climate Change as partial aspect of the ecological pillar of *Sustainable Development (SD)* into the UN *Sustainable Development Goals (SDGs)* links Climate Change directly to SD which might be defined as a development preserving existing essential items, systems, and values while adapting to new conditions in a flexible way under consideration of ecological, social, and economic aspects (Schwarz-Herion 2015a).

The SDGs are based on the *Rio + 20* outcome document *The Future We Want*. This document had set out mandates "...to establish an Open Working Group to develop a set of Sustainable Development Goals for consideration and appropriate action by the General Assembly at its 68th session..." (UN 2017b) and to ensure that the SDGs would be "...coherent with and integrated into the UN development agenda beyond 2015..." (UN 2017b).

The IPCC's reported findings include an average global temperature increase by 0.85 °C from 1880 to 2012, resulting into a decline in grain yield of maize, wheat, and other significant crops by 40 megatons per year from 1981 to 2002, an expansion of the oceans with a rise of the global average sea level by 19 cm due to warming and melting ice, and a shrink of the Arctic sea with a 1.07 million km^2 of ice loss per decade since 1979 (UN 2017a, with further references). According to the IPCC, "...it is likely that by the end of this century, the increase in global temperature will exceed 1.5°C..." (UN 2017a, with further references) as compared to the period from 1850 to 1900, the average sea level will rise by 24–30 cm in 2065 and 40–63 cm in 2100 due to the current concentrations and continued emissions of greenhouse gases, and "...most aspects of climate change will persist for many centuries even if the emissions are stopped" (UN 2017a, with further references).

Furthermore, the IPCC states that there were an increase of global CO_2 emissions by almost 50% since 1990 and a faster growth of emissions between 2000 and 2010

[1] United Nations Framework Convention on Climate Change.

than in any of the previous decades. The IPCC claims that a range of technological measures along with a "major institutional and technological change" (UN 2017a) will significantly increase the chance that "global warming will not exceed this threshold" (UN 2017a). So, the IPCC which is currently in its Sixth Assessment cycle (IPCC 2017a) expressly encourages technological innovations along with political reforms to deal with Climate Change.

Although the IPCC report cites concrete numbers and figures, the word "likely" regarding a certain temperature increase at the end of this century shows that the extent of Climate Change could not yet be fully clarified according to the IPCC. Other aspects of Climate Change remain equally controversial among scientists and scholars (PPO 2017a; Costella 2010). This includes the question if Climate Change is mainly due to natural or to human factors; the question whether mainly CO_2 and other greenhouse gases from **inadvertent** human activities like burning fossil fuels for industry, households, and street traffic or **deliberate** human activities like covert targeted weather modification and climate engineering are the main causes of Climate Change; as well as the question in how far anthropogenic (human-made) Climate Change is responsible for weather disasters and extreme weather patterns like heat waves, droughts, and frost (PPO 2017a; Costella 2010; Spencer 2007). Repeated claims of a "97% consensus" concerning supposed findings for certain Climate Change causes (Cook et al. 2013; NASA 2017) have triggered serious criticism (Ritchie 2016; Tuttle 2015).

Science is always in motion. An alleged "consensus" about any scientific finding is problematic and often untenable – especially if there are strong indicators that industrial lobbyists, covert political transformers, and influential church leaders try to shape people's opinion about certain topics. Climate Change can only be tackled successfully if many different opinions of unbiased scientists and scholars are thoroughly researched and logical conclusions are drawn from their findings to figure out the main causes of Climate Change.

Especially the true origin and the further development of the mono-causal *Climate Change-by-CO_2-hypothesis* in the period after the Second World War are still basically unknown to the general public. Since information and enlightenment are essential for an objective scientific discussion on Climate Change, the chapter at hand is supposed to fill this gap.

1.2 *The Limits to Growth* and Its Impact on Ecology, Economy, Politics, and Science

In 1972, a scientific report to the *Club of Rome*[2] *(COR)* entitled *The Limits to Growth* (Meadows et al. 1972) did not only discuss potential limits of further exponential economic growth, exponential population growth, industrialization, environmental

[2] Detailed information on the Club of Rome can be found on its website (Club of Rome 2017a).

pollution, food production, and exploitation of natural resources but also did the following startling prediction:

> At present about 97 percent of mankind's industrial energy production comes from fossil fuels (coal, oil, and natural gas). When these fuels are burned, they release, among other substances, carbon dioxide (CO_2) in the atmosphere…the measured amount of CO_2 is increasing exponentially apparently at a rate of about 0.2 percent per year…If the energy source is something other than incident solar energy (e.g. fossil fuels or atomic energy) that heat will result in warming the atmosphere….(Meadows et al. 1972 with references to Machta 1971; UN Department of Economic and Social affairs 1970; Bolin 1970; Inadvertent Climate Modification 1971)

This statement was based on observations of atmospheric concentrations of CO_2 at Mauna Loa, Hawaii, in 1958 which had reportedly increased steadily and averaged ca. 1.5 parts per million (ppm) each year. Calculations considering the well-known exchanges of CO_2 between atmosphere, biosphere, and the oceans predicted that the CO_2 concentration would reach 380 ppm by the year 2000, forming an increase of almost 30% of the supposed value in 1860. This "exponential increase in atmospheric CO_2" (Meadows et al. 1972 with reference to Machta 1971) was attributed to man's growing combustion of fossil fuels (Meadows et al. 1972 with reference to Machta 1971).

So, anthropogenic warming by industrial CO_2 emissions was already addressed in the early 1970s as a side aspect of this report which focused on the limits of economic growth and population growth due to reportedly increasingly scarce resources like food and fossil energies (Meadows et al. 1972). Written for the Club of Rome (COR) by an MIT research team under Dennis and Donella Meadows, *The Limits to Growth* was widely disseminated on a global base – with 12 million copies and translations into 37 languages (Suter 1999). Nonetheless, back then, *The Limits to Growth* was mainly embraced by environmentalists, whereas most politicians, managers, and economists showed little interest in it (Colombo 1997); some scientists and scholars even openly criticized it (Solow 1973; Shubik 1972; Kaysen 1972).

Already MIT scientist Jay Forrester's *World Dynamics* model (*World2*) as precursor model of the *World3* model on which *The Limits to Growth* was based had been criticized because of the model description of the world as a nonlinear feedback system, the application of computer modeling for social developments, and the infringement of scientific approach in the absence of empirical verifiability of the validity of such models (Shubik 1972). *The Limits to Growth* was mainly criticized by economists who considered the demand for a general state of equilibrium as unrealistic, arguing that the limitation of the raw materials was not a problem, since humans had always been able to adapt to resource constraints. *The Limits to Growth* was also criticized and ridiculed for its pessimistic view of the world (Solow 1973; Kaysen 1972).

In October 1973, however, the emerging global oil crisis seemed to support the report's message that excessive exploitation of nonrenewable resources would lead to serious problems (Colombo 1997). Nevertheless, the supposed economic necessity of the sudden explosion of the oil price due to the alleged shortage of oil supply would be called into question almost three decades later: In January 2001, *Ahmed Zaki Yamani* who had been the oil minister of Saudi Arabia from 1962 to 1986 told *The Observer* that the *Shah of Iran* had exposed *Henry Kissinger* as the driving force behind the increased oil price (The Observer 2001). Yamani also revealed that recently emerged documents from a secret conference proved that some British and US American state employees had been behind the orchestration of the increase of the oil price by 400% in the 1970s (The Observer 2001) – facts which had been completely unknown to the general public in the 1970s. Back then, *The Limits to Growth* and the subsequent oil crisis had led to a kind of win-win situation for different stakeholders: the environmental movement felt encouraged by the message that oil was a limited and environmentally harmful resource. Paradoxically, oil giants like the Rockefeller-owned *Exxon* equally benefited from the report's message that fossil fuels were a rare resource, because it provided a credible pretext for the significant increase in oil prices in the years following the publication of *The Limits to Growth.*

Henry Kissinger as string-puller behind the oil crisis of the 1970s had also played a major role in the Conference at which the authors of the book *Conditions of World Order* met in the Villa Serbelloni in Bellagio, Italy, from June 12 to 19, 1965, "… thanks to the hospitality of the Rockefeller Foundation" (Hoffmann et al. 1968), i.e., at the place where, according to some historians, also the foundation stone for the COR would be laid three years later in 1968 (Hap 2013; Rivera 1994), whereas the COR itself cites Rome as its official founding place – without mentioning any specific location (COR 2017a).

Allegedly, *Conditions of World Order* provided the base for *The Limits to Growth* (Hap 2013; Rivera 1994), although – except from the world food problem and the development of a new world model with the demand for a global government policy – at first glance, this book seems to have little in common with *The Limits to Growth.* Its possible connection to the topics of Climate Change and SD will be discussed in the final section of this chapter.

1.3 The Long-Term Impact of *The Limits to Growth* on International Climate Policy: From the First World Climate Conference in Geneva to COP21 in Paris and Beyond

The Limits to Growth (Meadows et al. 1972) is still listed as the most important report on the COR's website (COR 2017a), although it was succeeded by many other reports from 1975 to 2015 (COR 2017b). Six of them were published in the 1970s (COR 2017b):

- Mankind at the Turning Point (1975)
- Reshaping the International Order[3] (1976)
- Goals for Mankind (1977)
- Beyond the Age of Waste (1978)
- Energy: The Countdown (1979)
- No Limits to Learning (1979)

[3] This title indeed looks like an allusion to the book *Conditions of World Order.*

Perhaps inspired by the aforementioned COR reports, the late 1970s marked the start of a long series of international environmental conferences, especially Climate Conferences, finally leading to the COP21 Conference in Paris in 2015 and the Paris Agreement (UNFCCC 2017a), followed by the *22nd Conference of the Parties (COP22)* in Marrakech, Morocco (UNFCCC 2017b).

This series of international environmental conferences can be traced back to the *First World Climate Conference in Geneva* (February 12–23, 1979) at the invitation of the *World Meteorological Organization (WMO)*. Climate experts warned that the emission of greenhouse gases in the atmosphere could cause significant changes in the regional or even global climate, having a negative impact on the welfare of mankind. The *World Climate Research Programme (WCRP)* was established to further deepen the knowledge in this field (Staud 2015).

Nearly ten years later – on December 6, 1988 – the *Intergovernmental Panel on Climate Change (IPCC)* was established, based on *UN Resolution 43/53* (IPCC 2017a; Staud 2015) and sponsored by WMO and the *United Nations Environment Programme (UNEP)*. Every 6 years, experts from all over the globe produce their reports independently in three working groups; the short summaries are supposed to influence the governments of the 195 member states (IPCC 2017b; Staud 2015).

11 years after the launch of the *WCRP* and two years after the *IPCC First Assessment Report*, over 1000 experts and government representatives met in Geneva (October 29, 1990, to November 7, 1990) for the *Second World Climate Conference* (UNFCCC 2017c) where Margaret Thatcher stated: "The later we become active against climate change, the more expensive it becomes" (Staud 2015). Six weeks later, the UN General Assembly decided to start negotiations on a global climate deal (UNFCCC 2017c; Staud 2015).

At the *UN Conference on Environment and Development (UNCED)* in *Rio de Janeiro, Brazil,* June 3–14, 1992, the *Agenda 21*, the *Rio Declaration on Environment and Development,* and the *Statement of Principles for the Sustainable Management of Forests* were adopted by more than 178 governments (UN 2017c). On June 5, 1992, the *Convention on Biological Diversity (CBD)*, based on the controversial *Wildland Project,* which – with its core reserves (wilderness areas), corridors, and buffer zones – involves the risk of nationalization of private property in these areas, while shutting down half of the agriculture and reducing biodiversity instead of protecting it as revealed by official environmental statements (Coffman 2009), was opened for signature which would finally enter into force on December 29, 1993 (CBD 2017).

The *Agenda 21*, "...a comprehensive plan of action to be taken globally, nationally and locally by organizations of the United Nations System, Governments, and Major Groups in every area in which human impacts on the environment..." (UN 2017c), has long-term effects on environmental programs at the local level (Lexikon der Nachhaltigkeit 2015). In December 1992, the *Commission on Sustainable Development (CSD)* was created to supervise and report about the implementation of the agreements at the local, national, regional, and international levels. Additionally, a 5-year review of the Earth Summit progress scheduled for 1997 was decided in a special session of the UN General Assembly (UN 2017d).

In 1994, 155 countries signed the *United Nations Framework Convention on Climate Change (UNFCCC)*. It entered into force on March 21, 1994. The aim of Article 2 was "...to achieve the stabilization of greenhouse gas concentrations in the atmosphere at a level where a dangerous anthropogenic disturbance of the climate system is prevented" (Staud 2015).

From March 28 to April 7, 1995, the *First UN Climate Summit* took place in Berlin, where the signatory states of the Climate Convention met for the *First Conference of the Parties (COP1)*, led by *Angela Merkel* as the then *Minister of the Environment* of the host country Germany. Such *UN Climate Summits* were announced to be held annually. COP1 stated that state plans to reduce greenhouse gas emissions were **not** enough, mandating a working group to negotiate an additional protocol to the UNFCCC with **legally binding** emission reductions. It was also decided to establish the *Secretariat of the Climate Convention* in Bonn (Staud 2015). Since then, COP meets every year to take decisions that further the implementation of the Convention (UNFCCC 2017a, d).

From June 23 to 27, 1997, the *Rio + 5 Conference* took place in New York (UN 2017e). On December 11, 1997, at the *Third UN Climate Summit (COP3)* in Kyoto, Japan, the *Kyoto Protocol* as the **first legally binding agreement** on emission reduction was adopted. The EU with its then 15 members and further 23 industrialized countries committed themselves to reducing CO_2 and other greenhouse gases by an average of 5.2% by 2012. There were no similar requirements for developing countries because they hardly contributed to Climate Change. Germany undertook to reduce emissions by 21%. Although US President *Bill Clinton* signed the *Kyoto Protocol*, the *Senate* never ratified it (Staud 2015).

From July 16 to 27, 2001, there was a breakthrough at *COP6-2* in Bonn, an extension of *COP6* which had failed at the end of 2000 in The Hague, Netherlands. Back then, many states were slowing down, although the Third IPCC Assessment Report from 2001 warned that – until the end of the century – the earth could warm up to 5.8 °C. In early 2001, the newly elected President *George W. Bush* had declared the farewell to Kyoto for the United States. Finally, an agreement regarding the Kyoto Protocol was reached at COP6-2 but without the United States (Staud 2015).

From October 29 to November 10, 2001, at *COP7* in Marrakech, the compromises on the implementation of the Kyoto Protocol from *COP6-2* were described in the *Marrakech Accords*. The signatories were now entitled to claim forests and soils as CO_2 storage to compensate emissions. Industrialized countries could reduce emissions through projects abroad, so called "Joint Implementation" or "Clean Development Mechanism" (Staud 2015).

The *World Summit on Sustainable Development (WSSD)* in *Johannesburg*, South Africa, took place from August 26 to September 4, 2002. The full implementation of *Agenda 21*, the *Program for Further Implementation of Agenda 21*, and the commitments to the Rio principles were expressly confirmed (UN 2017c).

On February 16, 2005, the Kyoto Protocol entered into force. Over 100 states (by far more than the minimum requirement) ratified the agreement, but only the respective decision of the *Duma* in Russia at the end of 2004 led to the fulfillment of the second condition: the ratification states became responsible for 55% of the emissions, which had to be reduced according to the Kyoto Protocol. The EU launched a trading system for CO_2 emission allowances in 2005 to meet its obligations (Staud 2015).

At *COP13* (December 3–14, 2007) on the Indonesian island of *Bali*, the timetable for a Kyoto successor protocol was adopted. As the protocol would have expired in 2012, a connection agreement was negotiated under high pressure: an alliance between the EU and developing countries or emerging countries was able to make the United States abandon its blockade. In 2007, the IPCC received the Nobel Prize for its *Fourth Assessment Report* (Staud 2015).

In December 2009, the "ClimateGate" revelations about (suspected) manipulations of climate models irritated the scientific world and the general public (Booker 2009): *COP15* in *Copenhagen* (December 7–18, 2009), into which higher expectations had been placed than in any other Climate Summit before and where a connecting accord for the Kyoto Protocol had been planned, ended up with the "failure of Copenhagen" (Staud 2015). While the heads of state and the governments of many important countries negotiated every single word, other states denied their formal approval; the *Copenhagen Accord* remained a non-binding accord. The industrialized countries agreed to help the developing countries in climate protection and adaptation to the consequences of Climate Change by an annual contribution of 100 billion euros (Staud 2015).

After the failure of Copenhagen, the climate studies advanced in small steps. Although no agreement had been reached on a follow-up to the Kyoto Protocol, a restart at *COP16* (November 29 to December 10, 2010) in *Cancún* took place where the participating states explicitly decided for the first time in Climate Conference history to limit global warming to a maximum of 2 °C as compared to the preindustrial level (with the option to lower the limit later on to 1.5 °C). A *Green Climate Fund* was set up to manage the money which had been promised for the developing countries in Copenhagen. The participants of *COP16* adopted measures to protect the forests (Staud 2015).

At *COP17* in *Durban*, South Africa (November 28 to December 11, 2011), China and the United States made a move. The *Durban Platform*, adopted at *COP17,* brought about some progress: the United States agreed to the goal of a new, legally binding Climate Treaty. And for the first time, developing and emerging countries such as China generally agreed to abide to emission restrictions; the new contract should enter into force in 2020, but no agreement on how to proceed had been done. Nevertheless, the countries of the world finally decided to reach an agreement with *COP21* in 2015 (Staud 2015).

At *Cop18 (Kyoto II)*, November 26 to December 7, 2012, in *Doha (Qatar)*, the Kyoto Protocol has been extended to fill the gap to a new climate contract scheduled for 2020: the EU with its 27 member states and 10 other industrialized countries agreed to a "second commitment period." All in all, their target was to reduce their greenhouse gas emissions by 18% as compared to 1990 levels. However, as Japan, Canada, Russia, and the United States refused, and the Kyoto Protocol did not provide any obligations for emerging countries such as Brazil, China, or India anyway, the agreement concerned only 15 percent of global emissions (Staud 2015).

At *COP20* (December 1–14, 2014) in *Lima*, the delegates agreed on a first text draft for the new climate agreement to be decided in Paris in 2015. Now, in addition to the industrialized countries, the emerging and developing countries should also be obliged. By spring, all states were expected to report their targets to the *UN Climate Secretariat*, but according to expert estimates, they were far from reaching the goal of keeping global warming below 2 °C. Further commitments by industrialized countries and developing countries filled the *Green Climate Fund* with a total of 10 billion US dollars (Staud 2015).

The *Paris Climate Conference* was announced on the UN website as "the 21st Conference of the Parties" (*COP21*) to the *UNFCCC*. COP21 was scheduled simultaneously to *CMP11*, the *11th meeting of the Parties to the Kyoto Protocol*, which supervises the implementation of the Kyoto Protocol and the decisions taken to enhance its effectiveness (UN 2017f).

COP21 in Paris took place from November 30 to December 12, 2015. After a 4-year period of tough negotiations, the Parties to the UNFCCC reached a *landmark agreement* by ending the strict differentiation between developed and developing countries, replacing it with a common framework that committed all countries to do their best in the future (C2ES 2017).

Key outcomes of this conference were the Paris Agreement from November 2016 and a companion decision by the Parties which included reaffirming the objective of limiting global temperature increase "...well below 2 degrees Celsius, while urging efforts to limit the increase to 1.5 degrees..." (C2ES 2015), to set up "...binding commitments by all Parties to make 'nationally determined contributions' (NDCs) and to pursue domestic measures aimed at achieving them..." (C2ES 2015), to oblige "...all countries to report regularly on their emissions and 'progress made in implementing and achieving' their NDCs and to undergo international review..." (Raman 2016; C2ES 2015), and to "…extend the current goal of mobilizing $100 billion a year in support by 2020 through 2025, with a new, higher goal to be set for the period after 2025 and extend a mechanism to address 'loss and damage' resulting from climate change, which explicitly will not involve or provide a basis for any liability…" (C2ES 2015).

From November 7 to 18, 2016, *COP22* took place in *Marrakech*, *Morocco*, as a message to the world that "...the implementation of the Paris Agreement is underway and the constructive spirit of multilateral cooperation on climate change continues..." (UNFCCC 2017b). COP22 was also "...the twelfth session of the Conference of the Parties serving as the meeting of the Parties to the Kyoto Protocol (CMP 12), and the first session of the Conference of the Parties serving as the meeting of the Parties to the Paris Agreement (CMA 1)..." (UNFCCC 2017b).

On June 1, 2017, US President *Donald Trump* said in a speech that "…the United States will cease all implementation of the non-binding Paris Accord and the draconian financial and economic burdens the agreement imposes..." (Trump 2017), thus "…ending the implementation of the nationally determined contribution and…the Green Climate Fund which is costing the United States a vast fortune…this agreement is less about the climate and more about other countries gaining a financial advantage over the United States..." (Trump 2017). Trump pointed out that "…even if the Paris Agreement were implemented in full, with total compliance from all nations, it is estimated it would only produce a two-tenths of one degree…Celsius reduction in global temperature by the year 2100" (Trump 2017) and quoted a Wall Street article: "The reality is that withdrawing is in America's economic interest and won't matter much to the climate" (Trump 2017, with further references).

1.4 The Development of the Climate Change Discussion from 1988 to 2017: The Impact of Scientific Controversies on the Perception of Anthropogenic Climate Change

After the report *The Limits to Growth* (see Sect. 1.2), the next "scientific milestone" which got a similar level of attention from scientists and some interest from the general public were statements by James E. Hansen from *NASA* in *The New York Times* in June 1988: Hansen testified before a Congressional Committee that "…it was 99 percent certain that the warming trend was not a natural variation but was caused by a buildup of carbon dioxide and other artificial gases in the atmosphere…" (Shabecoff 1988), wrongly implying that CO_2 was allegedly an "artificial gas," although it is actually a naturally occurring gas in the atmosphere (Gettelman and Rood 2016; Schwarz-Herion 2015b, with further references).

Hansen was also quoted with the following statements: "…'It is time to stop waffling so much and say that the evidence is pretty strong that the greenhouse effect is here…Global warming has reached a level such that we can ascribe with a high degree of confidence a cause and effect relationship between the greenhouse effect and observed warming…It is already happening now.' …" (Shabecoff 1988). Hansen and other scientists testifying before the *Senate Panel* in June 1988 cited mathematical models predicting since some years that a "...buildup of carbon dioxide from the burning of fossil fuels such as coal and oil and other gases emitted by human activities into the atmosphere would cause the earth's surface to warm by trapping infrared radiation from the sun, turning the entire earth into a kind of greenhouse..." (Shabecoff 1988).

Six months after the publication of the aforementioned article, the IPCC was established (see Sect. 1.3). Although the IPCC had become an established authority in the field of Climate Change soon after its creation in December 1988, the Climate Change topic stays as controversial as the panel itself.

In 1999, a petition was signed by 31,487 American scientists – among them 9029 with a PhD in natural sciences or engineering – "…urging the United states Government to reject the global warming agreement that was written in Kyoto, Japan, in December 1997, and any other similar proposals…" (PPO 2017a) by the argument that the proposed limits on greenhouse gases would "…harm the environment, hinder the advance of science and technology, and damage the health and welfare of mankind…" (PPO 2017a). The scientists expressed doubts about the existence of hard scientific evidence that "...human release of carbon dioxide, methane, or other greenhouse gases is causing..." (PPO 2017a) or "...will cause catastrophic heating in the earth's climate and disruption in the earth's climate..." (PPO 2017a) in the near future, while highlighting scientific evidence for the many benefits of high atmospheric amounts of CO_2 on plants and animals (PPO 2017a). The signatories, listed with names and university degrees (PPO 2017d), refuted the allegations of "...'settled science' and an overwhelming 'consensus' in favor of the hypothesis of human-caused global warming and consequent climatological damage..." (PPO 2017c). They questioned claims of "...publicists at the United Nations, Mr. Al Gore, and their supporters..." (PPO 2017c) that only a few "skeptics" still doubted the catastrophic anthropogenic global warming development (PPO 2017c), stating that the anthropogenic global warming hypothesis was "...without scientific validity..." (PPO 2017a) and that "...government action on the basis of this hypothesis would unnecessarily and counterproductively damage both human prosperity and the natural environment of the Earth..." (PPO 2017a).

The methodology of the IPCC was equally criticized. The signatories challenged the UN's claims that the IPCC's research review was authored by nearly 600 scientists, arguing that these so-called "authors" were **only** allowed to comment on the **draft text**; the final version would neither correspond to nor include many of their comments, but would conform to the UN's objective in favor of "...world taxation and rationing of industrially-useful energy..." (PPO 2017b). Along with the petition, the scientists provided a representative review article encompassing 12 pages as a positive example for a transparent and serious scientific procedure, explaining that review articles would regularly ***not*** present new discoveries and that "...essential facts given in the review must be referenced to the peer-reviewed scientific research literature, so that the reader can check the assertions and conclusions of the article and obtain more detailed information about aspects that interest him..." (PPO 2017b).

In any case, the writer of this chapter witnessed that – in the period from 1999 to 2005 – most professors and students in environmental sciences stated that the anthropogenic Climate Change-by-greenhouse gases-hypothesis was anything but settled and had to be examined further. This was in contrast to sensational statements like the one of the UK's chief scientific advisor David King who tagged Climate Change as "...more dangerous than international terrorism" (Schwarz-Herion 2005, with further references).

Nevertheless, in 2006/2007, Nicholas Stern, a former *World Bank* economist and then economic advisor of the British Chancellor of the Exchequer, impressed the scientific world with the startling claim that the *gross national product* (GNP) would be reduced by 5-20% or even more if politics were unwilling to take measures against global warming, warning that this would trigger a global depression which would be even more dramatic than the *Word Economy Crisis of the 1930s*. This statement had been based on a study conducted by Stern for the British government in which he calculated the financial consequences of climate change for the world economy (Stern 2007). Soon after the *Stern review* came out, the support of the **mono-causal** anthropogenic Climate Change-by-greenhouse gases-hypothesis grew stronger in the academic world and in society as a whole. This might have been one of the reasons why the 1999 Petition Project was renewed in 2007.

The election of Obama in 2008 strengthened the supporters of the anthropogenic Climate Change hypothesis on a global base. Soon after his election, Obama confirmed his "...campaign vow to reduce climate-altering carbon dioxide emissions by 80 percent by 2050, and invest $150 billion in new energy-saving technologies..." when he spoke by video to a climate conference in Los Angeles. "...Now is the time to confront this challenge once and for all…Delay is no longer an option. Denial is no longer an acceptable response..." (Broder 2008).

One year later, however, the scientific world and the general public were shaken by the *ClimateGate* revelations: on November 19, 2009, thousands of emails and documents from the University of East Anglia's *Climatic Research Unit (CRU)* giving the impression of scientific fraud emerged (Costella 2010; Booker 2009). The media got wind of it very quickly. *The Guardian* articles, entitled "Climate Skeptics Claim Leaked Emails Are Evidence of Collusion Among Scientists" (Hickman and Randerson 2009) and "Pretending the Climate Email Leak Isn't a Crisis Won't Make It Go Away" (Monbiot 2009), stated that "...opaqueness and secrecy are the enemies of science…There is a word for the apparent repeated attempts to prevent disclosure revealed in these emails: unscientific" (Monbiot 2009).

Contradictory information about "leaking" via a whistle-blower versus "hacking" triggered further controversies. *The Wall Street Journal* headlined "Climate Emails Stoke Debate. Scientists' Leaked Correspondence Illustrates Bitter Feud over Global Warming" (Johnson 2009), whereas the NY Times headlined "Hacked Email Is New Fodder for Climate Dispute" (Ravkin 2009). Consequently, *The Guardian* asked: "Climate emails: were they really hacked or just sitting in cyberspace? Slack security or subversion at the university may have led to 'unintentional sharing', making the police investigation pointless", revealing that Climate email-"hackers" had "...access for over a month..." (The Guardian 2010).

A week after the scandal broke and the term "ClimateGate" had appeared on James Delingpole's *Telegraph* blog (Booker 2009), *The Telegraph* headlined "Climate Change: This Is the Worst Scientific Scandal of Our Generation" (Booker 2009), exposed "…the small group of scientists driving the worldwide alarm over global warming …", and stated that CRU's director *Philip Jones* with his connections to the *Hadley Centre* as part of the *UK Met Office* selected "…the majority of the IPCC's key scientific contributors, making his global temperature record the most important of the four sets of temperature data on which the IPCC and governments rely…for their predictions that the world will warm to catastrophic levels unless trillions of dollars are spent to avert it…" (Booker 2009).

The Telegraph called Jones "…a key part of the closely knit group of American and British scientists responsible for promoting that picture of world temperatures conveyed by Michael Mann's 'hockey stick' graph which 10 years ago turned climate history on its head by showing that, after 1,000 years of decline, global temperatures have recently shot up to their highest level in recorded history…the central icon of the entire man-made global warming movement" (Booker 2009), pointing out that the statistician Steve McIntyre had identified fundamental flaws in the "hockey stick" graph back in 2003 (Booker 2009). The leaked CRU emails involved influential personalities like Gavin Schmidt as "right-hand man to Al Gore's ally Dr

James Hansen" (Booker 2009). *The Telegraph* accused the scientists of deliberately having concealed the data on which their findings were based, exposing their claim of data loss as untrue: actually, the leaked emails showed that scientists had been actively advised to delete huge amounts of data after receipt of a freedom of information request, thus committing a criminal offense (Booker 2009).

The Telegraph criticized the use of computer programs, designed "…to lower past temperatures and to 'adjust' recent temperatures upwards…to convey the impression of an accelerated warming…" (Booker 2009). The newspaper exposed "…the ruthless way in which these academics have been determined to silence any expert questioning of the findings they have arrived at by such dubious methods… discrediting and freezing out any scientific journal which dares to publish their critics' work…" (Booker 2009) to keep any dissenting research out of the IPCC reports (Booker 2009). Finally, *The Telegraph* suspected that the inquiry into the CRU leaks requested by former Chancellor *Lord Nigel Lawson* and to be chaired by *Lord Rees*, the President of the *Royal Society*, a "shameless propagandist for the warmist cause" (Booker 2009), would **not** be conducted in an objective way and warned: "Our hopelessly compromised scientific establishment cannot be allowed to get away with a whitewash of what has become the greatest scientific scandal of our age" (Booker 2009).

A representative selection of the leaked (hacked?) email correspondences was edited and annotated in a scientific way in a publication entitled *The ClimateGate Emails* by the Australian *John Costella*, a trustworty judge of data reliability, who holds a PhD in physics and had been in charge of analyzing statistical data as data reliability engineer at the *Department of Defence* in 2006. This publication supports all accusations addressed in *The Telegraph* article. Furthermore, Costella provides many other details such as the key players' *Modus Operandi* (M.O.), e.g., "inventing" the 1995 winter temperatures; discussing machinations to influence the delegates in advance of the *Kyoto Protocol*, cabals to harass and silence critics like members of the former editorial board of the scientific journal *Climate Research*; or cunning ways to get scientific results which please the EU and to receive funds from *Shell* for their research (Costella 2010).

The ClimateGate scandal kept the media busy in the years following the scandal. In 2010, a *Spiegel Online* article headlined "Forscherskandal: Heißer Krieg ums Klima" ("Researcher's Scandal: Hot War on the Climate"). *Spiegel's* scientific journalist Bojanowski stated that climate researchers had gone into the trap of the industrial lobby. Having read and analyzed the over 1000 leaked ClimateGate emails, freely available on the Internet and covering a period of 15 years, *Spiegel* found out that leading researchers had become "…entangled into a fierce and serious trench war, partly by fierce external attacks, including media, environmental associations and politicians" (Bojanowski 2010). Stating that "uncertainties of the results of the research will probably continue to exist in climatology," *Spiegel* concluded: "Politicians should not listen to scientists who promise simple answers" (Bojanowski 2010).

After the temporary outburst of indignation in the scientific world and the media from 2009 to 2011, the ClimateGate affair was silenced rather quickly: by the end of July 2012, the *BBC* reported "Police end 'ClimateGate' inquiry", mentioning that the *Norfolk Constabulary* saw no realistic prospect of finding the culprit responsible for the (supposed) hacking but also mentioned that the examination of the broader context had resulted in the scientists being cleared of malpractice (Black 2012).

1.5 Economic Aspects of the Climate Change Discussion and the Potential Economic Impact of the Paris Agreement

In 2015, *Nicholas Stern* published a book entitled *Why Are We Waiting? The Logic, Urgency, and Promise of Tackling Climate Change*. In this book, based on the reported findings of the IPCC from 2014 and on data from the *International Energy Agency* (IEA), Stern points out that "...the science of Climate Change should be the foundation both for an understanding of the issues and challenges and for any proposed responses to those challenges..." (Stern 2015). The main message of this book seems to be that the risks and costs of Climate Change are higher than Stern had supposed in his earlier review (see Sect. 1.4) and that technological innovation and international cooperation should be encouraged to reduce environmentally harmful emissions and to push economic growth at the same time (Stern 2015).

Stern appeals to the media to optimize the discussion on Climate Change: "The importance of frequent, accurate, clear, and accessible public discussion of climate change places a great responsibility on media organizations" (Stern 2015). He propagates the display of destruction by weather disasters like typhoons on TV to show the consequences of Climate Change and complains about a communication deficit regarding sound arguments in favor of Climate Change. Stern dedicates nearly 10% of his book to moral philosophy and political philosophy which did not appear in his earlier publications: "Climate change presents a range of normative moral and political questions that cannot be dodged" (Stern 2015).

In a report entitled *COP21 at Paris: The issues, the actors, and the road ahead on climate change*, a team of authors (mainly economists) – among them Nicholas Stern – had analyzed the expected effects of this Climate Summit just before COP21 at Paris (Bhattacharya et al. 2015). The authors stated that the governments from ca. 200 nations who would come together in Paris for COP21 had "high expectations", striving for a legally binding and universal agreement on reduction of global temperature increases, and that brooking experts had "…compiled a collection of comprehensive short briefs on key issues in climate action, including climate aid and finance, infrastructure, carbon pricing, the relationship between agriculture, and climate…" (Bhattacharya et al. 2015). They considered 2015 an especially successful year for "…climate change efforts, as the long-floundering U.N. process has finally begun to deliver some of what is needed" (Bhattacharya et al. 2015).

The authors predicted that COP21 would become "...an important turning point in the fight against climate change…" (Bhattacharya et al. 2015), underlining that all significant world leaders including the ones of China and the United States had declared their readiness for "...an ambitious Paris agreement..." (Bhattacharya et al. 2015). The authors praised the UNFCCC's efforts to advance a new agreement for the period after 2020, stressing the supposed importance of pricing carbon and encouraging more discussion on the "…most cost-effective means to reduce emissions", e.g., "…reducing fossil fuel subsidies and pricing greenhouse gas emissions…" **rather than** "…assistance to poor countries for adaptation and mitigation... clean energy cooperation, forest preservation…" (Bhattacharya et al. 2015).

Bhattacharya et al. promoted infrastructure investment as "…key element of the climate change agenda…" (Bhattacharya et al. 2015), alleging that infrastructure would account for more than half of global carbon emissions, praising the *2009 Copenhagen Accord* as "…the first effort to spell out the financial implications of a global effort to reduce carbon emissions where developed countries obliged themselves to provide 30 billion dollars for mitigation and adaptation financing for the period 2010-2012, to mobilize 100 billion dollars per year by 2020…" (Bhattacharya et al. 2015). While propagating climate finance, the authors vilified agriculture as "…one of the foremost drivers…of climate change" (Bhattacharya et al. 2015), stating that – regarding agriculture – "…the COP21 agreement might best be dubbed 'Les Champs-Oubliés', or forgotten fields…" (Bhattacharya et al. 2015).

In sum, this report seemed to be completely focused on the economic pillar of SD while neglecting the ecological pillar by failing to promote inexpensive energy-efficient technology and forest preservation which is essential to provide natural sinks for CO_2 (Schwarz-Herion 2005, with further references). As the climate in most cities is by far worse than the climate in the countryside, the authors' denigration of agriculture, implying that further urbanization was supposedly the better alternative, is also problematic for the ecological pillar. Considering the content of the Paris Agreement (see Sect. 1.3), equally focusing on the economic pillar of SD, the aforementioned report might have had some influence on this agreement.

Soon after the Paris Agreement, a publication entitled *Report Report COP21: So, COP21 happened – what next?* which appeared in February 2016 gives a rather clear picture which stakeholders benefit from COP21 and the Paris Agreement. This "report report" highlights the basic points from recent reports of business groups and analysts. The author mentions that Al Gore's *Climate Reality Project* had called the Paris Agreement a "turning point" which had sent "...a clear signal to markets and investors that the future of energy is in renewables such as wind and solar…" (Hower 2016), pointing out that "...a low-carbon economy would be a boon to renewables, energy storage and the IT sector, among others..." (Hower 2016).

First, Hower summarizes the *Bloomberg Energy Finance Report* which gives a forecast for a significant increase in the total investment into lower-carbon technologies over the decade: "The 2 degrees Celsius scenario represents a $12.1 trillion investment opportunity for new renewable electric power generation over 25 years, or $485 billion per year on average. A majority of anticipated investment in new renewable power generation is likely to go toward emerging markets…Non-OECD countries are expected to attract $4.3 trillion for new renewables generation through 2040…." (Hower 2016, with further references). Hower basically confirms this view, predicting a huge "...market opportunity for small-scale solar and wind technologies, microgrids and energy efficiency innovations" (Hower 2016) for regions without any access or only limited access to electricity, helping the 1.1 billion people who had no reliable access to electricity yet to "...alleviate some longstanding social and environmental issues" (Hower 2016).

Certainly, reliable access to electricity facilitating the regular use of computers and cell phones in favor of information equality for people all over the world is generally desirable. It is also reasonable to use solar and wind energy which are basically "clean" renewable energies, if used in the proper way and at the right places (Schwarz-Herion 2015b, with further references). Nonetheless, people from different nations and cultures might have different priorities. It is questionable if investment into infrastructure projects in favor of energy security is a priority for those communities of developing countries who currently live off-grid – especially if current and future generations might accumulate debt for expensive infrastructure projects as it happened in the past (Perkins 2005) while putting windmills on land which might be needed for food and facing problems with heaps of electronic waste in a scope previously unknown to them.

Hower also basically supports the judgment of the *Sustainalytics Report*. This report praises the Paris Agreement as a "Triumph of the Optimists" (Hower 2016, with further references), predicts that subsequent national climate commitments might benefit from this "…positive long-term economic signal for low-carbon technologies…", and states that the Paris Agreement would help to foster low-carbon technologies, although the end of fossil energies would still lie in the far future and that companies like *Tesla*, *Cisco*, *Kellogg*, and *L'Oréal* allowed investors to make full use of "…the regulatory, market and physical impacts of climate change…" (Hower 2016, with further references).

Hower equally basically verifies the predictions of the *Accenture Report*, that the coming decades will lead to "...shifting patterns of global growth – particularly in emerging economies – making customers a 'moving target'", while demographic changes create big challenges for consumer and labor markets. According to Hower, digital empowerment will raise consumers' expectations and lead to increased competition between companies while data companies will have a huge potential concerning data collection. Hower additionally stresses that increasingly scarce resources will make it difficult to cope with decades of unrestrained growth and unsustainable consumption patterns (Hower 2016) but is optimistic that digital technology will play a major role in the future of sustainable business by optimizing resource efficiency and companies' accountability for their actions and that information communication technologies might create annually more than $11 trillion by 2030, "…equivalent of China's annual GDP in 2015…" (Hower 2016, with further references).

Admittedly, increased digitalization facilitates sharing best-practice SD concepts digitally on a global base but can also cause serious social problems by reducing labor force even further. Additionally, the short life of technological devices used in the framework of steadily expanding digital markets (Soltan 2016a) will produce an unprecedented amount of "technotrash" and "e-waste", further adding to the estimated 20–50 million tons of technotrash already produced annually on a global base (Soltan 2016b) – ecological damages intrinsic to the *Fourth Industrial Revolution* (Ribeiro 2016).

According to the *Climate Policy Observer's* report *Carbon Market Monitor: America to the Rescue*, the carbon trading market showed two contrarotating trends in 2015: "Volumes continued to contract, and prices continued to increase" (Hower 2016, with further references). The North American markets showed growth in volume by 121% and growth in value by 220%, while Europe still remains the market leader (Hower 2016, with further references). Hower claims that the "...new growth of cap-and-trade in North American markets and continued growth in Europe..." were heartening and showed that "...climate change action won't stifle business growth", encouraging politics to strive for a low-carbon economy while keeping global temperatures at or below 2 °C in a "post-COP21 world" (Hower 2016).

So, influential business people seem to be very optimistic regarding the economic potential of Climate Change deals. This market optimism might be one of the reasons why the IPCC continues its work regardless of criticism.

1.6 Climate Change in the Media, Literature, and Cinema

Media coverage on anthropogenic Climate Change is suitable to shape public opinion and is largely responsible for the ways in which science is translated into policy regarding the environment, new technologies, and risks (Boykoff and Rajan 2007, with reference to Weingart et al. 2000). Since several years already, the mass media split the global public in different fractions: "climate supporters" or "climate alarmists" on the one hand and "climate skeptics" or "climate deniers" on the other hand (Wikipedia 2017; Wiki-Talk 2017). Although the topic of Climate Change has a huge media coverage in Europe, particularly in Germany, some complain that the impact of Climate Change would not get sufficient coverage in the US American media, since only few TV providers, e.g., PBS, featured "…stories on Pope Francis' environmental encyclical, the Paris Climate Summit, the Keystone Pipeline, and the Clean Power Plan..." (Yerman 2016).

Others make fun of the present notion considering the future as catastrophe – in cinema, science, and literature (Horn 2014). In "Zukunft als Katastrophe" ("Future as Catastrophe"), a "...scholarly, exciting essay on apocalypse presentations in literature and film..." (Hugendick 2014), the Germanist *Eva Horn* discusses the dangers of modern disaster awareness in literary and movie scenarios, enhancing fears of an apocalypse due to (alleged) overpopulation, providing the message that nature reconquers its space, depopulation and overpopulation scenarios leading to famine, occasionally coupled with cannibalism like in McCarthy's empty world in his post-apocalyptic novel *The Road*, the "compulsive denaturation of man" dreaming of his own extinction, and the risk of political instrumentalization, creating the impression that "…there must be no more compromises…that it has to be done, if there is still some time left to act…" (Horn 2014; Hugendick 2014).

This also goes for movies featuring Climate Change and weather disasters. Recently, *The Guardian* cited five movies as "the five best climate change films" (The Guardian 2017), reviewing them regarding ideological, environmental, and social messages as follows:

- In *Snowpiercer* (2013), directed by *Bong Joon-ho*, "…climate engineering has plunged the Earth into a new Ice Age, and the only survivors all inhabit a huge train that perpetually circumnavigates the globe…a microcosm of 21st-century society: multicultural but with 1% in charge of the engine…" (The Guardian 2017).
- *The Day After Tomorrow* (2004), directed by *Roland Emmerich*, "…his most clarion environmental statement...smuggles the eco-consciousness of the director's native Germany into the heart of Hollywood…visualizing the atmospheric consequences of the North Atlantic current slowing down and causing superstorms…Fans of unsubtle political statements could also thrill to the sight of southern American climate refugees streaming across the border to Mexico…" (The Guardian 2017).
- In *Hell (2011),* directed by *Tim Fehlbaum* and produced by Roland Emmerich, "…the Mad Max franchise takes place at the upper end of the thermometer's scale…A temperature rise has made daytime excursions dangerous…survivors trail distant birds in search of water…Hell mostly follows the post-apocalyptic playbook…The film's economic plotting…is a lesson in itself for a more resource-scarce world" (The Guardian 2017).
- *Beasts of the Southern Wild* (1995), featuring Kevin Costner, provides "...a different kind of rising sea picture…A six-year old Hush Puppy…living in a Louisiana bayou community called 'the bathtub' has a head full of premonitions of prehistoric aurochs that her teacher tells her will be released by melting ice caps. Their snouts and hooves represent a fear of ancient, resurgent nature… southern magical-realist fable, inspired by the Hurricane Katrina disaster…" (The Guardian 2017).
- In *Interstellar* (2014), directed by *Christopher Nolan*, "…giant dust storms causing crop blight that is destroying the global food supply…It's the last okra harvest ever, we're told. Everyone is in the last-chance saloon, scrabbling for basic resources…All the classic Midwest cornfield imagery isn't just Nolan's wake-up call to the most climate-change sceptic country, but a reminder to us all that any solution begins in the most familiar place: home" (The Guardian 2017).

The sudden Ice Age scenarios in the movies *Snowpiercer* and *The Day After Tomorrow* awaken associations to smoke-induced Climate Change from nuclear warfare as discussed in two *Science* papers from 1983, based on *Nuclear Winter* studies (Holbraad and Petersen 2013, with reference to Turco et al. 1983 and Ehrlich et al. 1983). Turco et al. used former models, originally developed "…to study the effects of volcanic eruptions…" (Turco et al. 1983), stating that nuclear war could trigger a major cooling of the Earth's atmosphere, the so-called "Nuclear Winter". Nuclear airburst of all yields would burn cities and forests; the dust and smoke would "…encircle the earth within 1 to 2 weeks…" (Turco et al. 1983) –

with the potential to reduce light levels to a minimum. The result would be subfreezing land temperatures ranging from −15 ° to −25 °C, even in summer (Turco et al. 1983). Ehrlich et al. cited studies of large-scale nuclear war (5000- to 10,000-MT yields), based on estimations that 750 million immediate deaths would result from blast alone. The combinations of the effects from blast, fire, and radiations might even cause 1.1 billion deaths along with ca. additional 1.1 billion injuries, erasing 30–50% of the total human population with the vast majority of fatalities in the Northern Hemisphere (Ehrlich et al. 1983).

In *The Day After Tomorrow*, equally the Northern Hemisphere is struck by the climate catastrophe, leading to the collapse of civilization: "Resonant images of a wall of water crashing down on Fifth Avenue in New York, the Empire State Building, and New York skyscrapers cracking apart, the Statue of Liberty half-buried in a frozen ice-block, and tornadoes ripping the letters of the Hollywood sign, produce an imagination of disaster that provides cautionary warnings about environmental breakdown" (Kellner 2009).

Although *The Day After Tomorrow* is not about nuclear war, it uses the "… devices of atomic cinema – focus on the destruction of cities, collective sacrifice and militarized response – to address a form of catastrophe larger than the national politics of the security state: radical climate change" (Holbraad and Petersen 2013). Its title equally seems to allude to the 1983 nuclear war film *The Day After*. Some movie analysts consider *The Day After Tomorrow* a "…loose remake of the 1961 feature *The Day the Earth caught Fire...*" (Holbraad and Petersen 2013) where an "…aggressive series of nuclear tests by the United States and Soviet Union knock the Earth off its axis, causing the planet to spin closer to the sun, and producing a nearly apocalyptic heat wave…" (Holbraad and Petersen 2013).

Emmerich seems to have known about the possibilities of Climate Change by weather manipulation decades before *The Day After Tomorrow*: his first movie *Noah's Ark Principle* (1984) had been a science fiction movie about weather control by the CIA (Film Atlas 2017; Maguire 2008) and came out just one year after the publication of the scientific papers about the *Nuclear Winter* studies mentioned above. The warlike scenarios in *The Day after Tomorrow* suggest that "…climatic affects would be as serious as war…" (Ehrlich et al. 1983), an aspect which was also addressed in a *Pentagon study* focusing on the national security implications of abrupt climate change, published a year before *The Day After Tomorrow* (Schwartz and Randell 2003).

Nevertheless, *The Day After Tomorrow* is outwardly based on the hypothesis that CO_2 emissions from fossil fuels were responsible for Climate Change. In the movie, paleoclimatologist Jack claims in a speech at a UN Conference in New Delhi that 10, 000 years ago, a global warming had changed the earth's climate and led to the Ice Age which might happen again in 100–1000 years if we wouldn't stop "polluting" the atmosphere: The melting of the polar ice caps would pour fresh water into the ocean, thus diluting the salt level and making the temperature of the ocean currents drop by 13 °C. This explanation was also provided in the aforementioned Pentagon study about Climate Change published in 2003 (Schwartz and Randell 2003; Schwarz-Herion 2005, with further references) which might have been one of the reasons why this movie had drawn many professors and students from environmental sciences into the cinemas back in 2004. In the movie, Jack's prediction is time-wise even surpassed by the real happenings with Climate Change-related weather disasters like high floods, tornados, huge hailstones destroying Tokyo, and snowstorms taking place in time-lapse of only some days.

Another well-known movie produced in the first decade of the New Millennium is *Al Gore*'s movie *An Inconvenient Truth* from 2006, a film documentary which "… aims to call attention to the dangers society faces from climate change, and suggests urgent actions that need to be taken immediately..." (Masters 2017). Although 50,000 free copies of the $19.99 DVD had been offered to the *National Science Teachers Association* (NSTA) for use of the film in US classrooms, the NSTA turned this offer down by the argument that the NSTA had a 2001 policy against "product endorsement" (Masters 2017).

In contrast, a recent, award-winning film documentary by Marijn Poels entitled "The Uncertainty Has Settled" shows the dangers of globalization along with climate politics causing radical changes such as farmers becoming energy suppliers and asks the question: "Are we doing the right thing?" (Sputnik 2017). Poels tries to increase public awareness about the negative effects of a too extensive and too rapid expansion of renewable energies, drawing people's attention to the fact that food supply is more important for the survival of humankind than energy supply, even if "green" energies are used (Sputnik 2017).

An over-exaggerated fear of CO_2 as possible main cause of Climate Change might produce similar results like Mao's *Great Sparrow Campaign* which marked the start of "…the greatest mass starvation in history…" (Platt 2013): in 1958, Mao ordered to kill all sparrows to protect the farms, declaring sparrows as pests because they ate grain. So, the Chinese killed hundreds of millions of sparrows. The consequences of this campaign became evident in 1960 because sparrows do not only eat grain seeds but also eat insects and – with the birds gone – insect populations boomed: locusts swarmed all over the country and ate everything they could find including crops. Scholars estimate that 45–87 million people died from starvation due to this environmentally and socially disastrous campaign (Platt 2013). Although Mao finally ordered an end of his sparrow-killing campaign, it took decades until the people, the environment, and the economy fully recovered (Pusa 2017).

This example illustrates what can happen if a generally useful naturally occurring thing, e.g., a living thing like useful animals or a naturally occurring useful gas like CO_2 is demonized. Misleading terms like "decarbonization" or "zero CO_2" have the potential to lure unsuspecting people into the erroneous belief that CO_2 was a noxious or even toxic gas which needed to be extremely reduced or even entirely eliminated. The application of these terms, however, has the general potential to cause a mass hysteria which might finally lead to a severe and long-lasting disturbance of the ecosystem, because CO_2 is "…not something 'foreign' to our bodies…" (Gettelman and Rood 2016), but "…part of the system, a critical part that is naturally all around us. We drink CO_2, we exhale it, and our bodies are made of carbon. Carbon is absorbed by plants with photosynthesis and used to build their tissues. So, CO_2 is not bad; it is a natural part of the system" (Gettelmann and Rood 2016). Thus, the term "decarbonization" would mean the end of all life on earth.

1.7 The Instrumentalization of the Climate Change Topic by Political Transformers and Its Potential Impacts on Global Politics and Society

Occasionally, parts of the media and the scientific world argue that only a global government along with a global police force might enforce "climate justice." Would this really be necessary for the greater good or might it rather open the door for abuse of power by a global authority and – in the worst case – possibly even lead to a global fascist dictatorship?

The suggestion to put a global policy in place had already been made in the comment of the COR's Executive Team on *The Limits to Growth* (King et al. 1972). King et al. state that:

> …the global issue of development is…closely interlinked with other global issues…The achievement of a harmonious state of global economic, social, and ecological equilibrium must be a joint venture based on joint conviction with benefits for all...the Club of Rome also will encourage the creation of a world forum where statesmen, policy-makers, and scientists can discuss the dangers and hopes for the future global system without the constraints of formal intergovernmental negotiation. (King et al. 1972)

This leads us back to *Conditions of World Order* as a possible precursor book for COR reports from the 1970s to 2015 which might have inspired the third book of this series, *Reshaping the International Order*, coordinated by Jan Tinbergen, one of the co-authors of *Conditions of World Order. Reshaping the International Order* puts a focus on "…development, distribution, and improved welfare that will require a good deal of economic growth…" (Tinbergen et al. 1976).

Conditions of World Order, influenced and co-authored by the controversial politician Henry Kissinger,[4] who has meanwhile been exposed by former counter-intelligence officers as an "avowed agent of British oligarchic interests... in a 'multi-polar-world'" (Thompson 1982) and as an active member of subversive supra-national elitist circles (van der Reijden 2017; Thompson 1982; Day 2012), propagates the supposed "indispensability of World Order" (Jaguaribe 1968). Kissinger, who would later on support *Operation Gladio* and similar operations financially and personally as revealed by declassified secret service papers and other credible sources (Fleming 2017; van der Reijden 2013), argues that "…the ability to foment domestic unrest is a more potent weapon than traditional arms" and that many country leaders "…will be very sensitive to the threat of domestic upheaval… all states find themselves face to face with the necessity of avoiding a nuclear holocaust…a common task which technology will impose…Therefore, we must establish an international order before a crisis imposes it as a necessity" (Kissinger 1968).

---

[4] Kissinger who is known for statements like "The illegal we do immediately; the unconstitutional takes a little longer…" (NSA Archive 2012) is also accused of having jeopardized US efforts to stop mass killings by Argentina's 1976–1983 military dictatorship by congratulating the country's military leaders for "wiping out" terrorism, as shown by a large trove of newly declassified state department files" (Goñi 2016; CIA 2017).

The authors of *Conditions of World Order* discuss possible chances to reach this objective, including the possibility of using UN as police force for the World Order (Hoffmann et al. 1968) and expressing their regret that "…the unofficial conference of NATO parliaments could not evolve into a NATO parliamentary assembly…" (Van Benthem Van Den Bergh 1968). This might have allowed the United States and Europe to share competences in the fields of economics and defense (Van Benthem Van Den Bergh 1968). Anyway, Hoffmann states that the "...European experiment" (Hoffmann 1968) basically succeeded as the European states survived transformed, "...swept by the advent of the 'age of mass consumption'...caught in an apparently inexorable process of industrialization, urbanization, they become more alike in social structure, in economic, and social policies..." (Hoffmann 1968), whereas Tinbergen states that the UN and their specialized agencies were "...closest to a World Organization..." (Tinbergen 1968) including the *UN Organization for Industrial Development (UNOID)*, the *Food and Agricultural Organization (FAO)*, the *International Bank for Reconstruction and Development (IBRD)*, the *International Monetary Fund (IMF)*, and the *World Health Organization (WHO)*, a concept supported by Henry Kissinger who is often quoted with the words: "Who controls the food controls the people...he who controls the money, rules the world" (Mittelbach 2013).

In the food sector, the FAO/WHO-established *Codex Alimentarius*, "...a collection of standards, guidelines and codes of practice...by the Codex Alimentarius Commission...to protect consumer health..." (FAO 2017) is nowadays quite influential, but criticized by independent health experts for "having to do with wealth, not health" (Health Freedom USA 2017) by supporting the pesticide-, GM-, nanotechnology-, and pharma-lobby (Todhunter 2017; Byrne 2007; Dr. Rath Foundation 2005) and using public interest groups, funded by social engineering groups like the *Rockefeller Family Fund*, who are demonizing essential vitamins by deeply flawed, "media-ballyhooed" studies which can easily be refuted by experts from, e.g., Harvard Medical School (Byrne 2005). In the financial sector, the politicization of the IMF, claiming to promote "global economic stability" by helping "...countries implement sound and appropriate policies..." (IMF 2014), triggers criticism as "...'countries...involved with IMF loans...all fail!' "while other countries gain political power over them via the IMF, "... peddling ... the agenda of its strongest member ... over the needs of those it claims to help..." (Matsangou 2017). Experts from Stanford stated already after the Asian financial crisis from 1997 that the IMF as a "lender of last resort" facilitates situations of "moral hazard" arising if "an actor does not bear all the risks of his actions", recommending that the IMF should "... focus on being a capital provider...not a policy dictator..." (Moore et al. 1999).

In sum, Tinbergen calls for "…a more active role of the United Nations…" in the future development of international economic planning, while criticizing the UN's allegedly "weak construction" (Tinbergen 1968) and its "strong preference for national autonomy" (Tinbergen 1968). Meanwhile, Tinbergen's concept seems to have been partly implemented on a global base - at least in the food, health, and financial sectors; the plans of the team of authors who contributed to *Conditions of World Order*, however, do not stop here but go far beyond that.

Helio Jaguaribe, nowadays still actively involved in social-democratic projects, e.g., participating in political film documentaries (Jaguaribe 2017), describes the envisioned future *World Order* as follows:

> …Only the Authority is Sovereign and only it can legislate for the world in general and provide for the observance of its norms. The national states, and not the individual citizens of these states, are the subjects of the Supra-National Authority. The organs of the Authority are established not directly by the citizens of the national states, but by the states themselves. Representation of each state will probably not be merely proportional to its population, but will take into consideration other factors, such as total and *per capita* income, previous military power…. (Jaguaribe 1968)

Bestseller authors Glenn Beck and Harriet Parke might have had Jaguaribe's World Order concept in mind when they wrote the postapocalyptic thriller *Agenda 21,* whose commercial summary gives a descriptive insight into the dire scenarios of this thriller:

> Just a generation ago, this place was called America. Now, after the worldwide implementation of a UN-led program called Agenda 21, it's simply known as "the Republic." There is no president. No Congress. No Supreme Court. No freedom. There are only the Authorities.
>
> Citizens have two primary goals in the new Republic: to create clean energy and to create new human life…This … is all that eighteen-year-old Emmeline has ever known…Until the day they come for her mother...Emmeline begins to search for the truth. Why are all citizens confined to ubiquitous concrete living spaces? Why are Compounds guarded by Gatekeepers who track all movements? Why are food, water and energy rationed so strictly? And…why are babies taken from their mothers at birth? As Emmeline begins to understand the true objectives of Agenda 21 she realizes that she is up against far more than she ever thought. With the Authorities closing in, and nowhere to run, Emmeline embarks on an audacious plan to save her family and expose the Republic—but is she already too late?" (Beck and Parke 2013)

The thriller *Agenda 21* strongly reminds the concept of a World Order as the one described in *Conditions of World Order*, although the mutual threat of nuclear power as "common task" or "unifying element" to justify a new international order (Kissinger 1968) has been replaced by the threat of Climate Change. In real life, the *Agenda 21* (see above Sect. 1.3) and the *2030 Agenda for Sustainable Development* adopted at the *United Nations Sustainable Development Summit* on September 25, 2015, a "new universal Agenda" (UN 2017g) which seeks "…to build on the Millennium Development Goals and complete what these did not achieve…" (UN 2017g) seem to have noble objectives, but both are widely criticized.

Some consider the *2030 Agenda for Sustainable Development* a "…blueprint for the global enslavement of humanity under the boot of corporate masters…" (Adams 2015). Others consider it a "…perfect vehicle for the elite…to bring in their version of utopia, because…every possible form of human activity affects the environment in some way…" (Snyder 2015). Apparently, "…U.N. Agenda 21 has morphed into Agenda 2030 and now into Vision 2050, a plan to force 9 billion people to live by the globalists prescription of 'living well and within the planet's resources…" (Johnson 2015). *Vision 2050* reaped ironical remarks in the German newspaper *Die Welt* in an article entitled "Bis 2050 wird der 'Normalbürger' abgeschafft," i.e., "Until 2050 the 'normal citizen' will be abolished" (Maxeiner and Miersch 2011).

Jørgen Randers, one of the co-authors of *The Limits to Growth*, has even visions beyond 2050: in a COR report entitled *2052 A Global Forecast for the Next Forty Years*, Randers envisions "…impressive advances in resource efficiency, and an increasing focus on human well-being rather than on per capita income growth..." (Randers 2012) but predicts that "…future growth in population and GDP…will be constrained…by rapid fertility decline as result of increased urbanization, productivity decline as a result of social unrest, and continuing poverty among the poorest 2 billion world citizens…" (Randers 2012) and that "…runaway global warming, too, is likely…" (Randers 2012). Randers states that "…the world is small and fragile, and humanity is huge, dangerous, and powerful…", calling his daughter "...the most dangerous animal in the world..." (Randers 2012) who "…consumes between 10-30 times as many resources and generates 10-30 times as much pollutants as an Indian child…", encouraging birth controls in countries of the industrialized world (Randers 2012) and suggesting that - due to the urgency of mitigating Climate Change by CO_2 emissions - democracy should be replaced by an authoritarian style of government. Randers predicts a future in which people will not know nature and animals anymore, will not be able to travel anymore and where unrest will be suppressed by fighting robots (Randers 2012).

In another recent COR report entitled *Reinventing Prosperity*, co-authored by Jørgen Randers and Graeme Maxton, the authors identify "…persistent unemployment, widening income inequality, and accelerating climate change..." as "…most challenging problems of our time" (Randers and Maxton 2016), offering 13 "…politically feasible proposals to improve our world…" (Randers and Maxton 2016) including shortening the work year and raising the retirement age. The authors reaped derision and mockery for this book. The German newspaper *Die Zeit* headlined "Club of Rome: Die Welt auf die harte Tour retten" ("Club of Rome: Saving the World the Hard Way") and wrote: "In the brand new report to the Club of Rome, which Randers presented together with his Secretary General Graeme Maxton in Berlin, the co-authors are even more concrete. Any woman who is raising only one child should get a $ 80,000 bonus at the age of 50 for her renunciation…Such a rigorous state attack into one of the most private decisions of every individual citizen, and with a slippery financial incentive: this is not only unpopular but also ethically questionable" (Grefe 2016).

Enlightened and morally responsible modern scholars warn against the "fatal misconception" to consider deliberate depopulation as key measure "...to prevent poverty, hunger, ecological decay, and runaway climate change" (Butler 2009, with further references) or even to equate it "with the very survival of humanity" as "...the hardcore populationist lobby" ignores the fact that "...population growth is slowing worldwide", treats "...the victims of social and economic injustice as obstacles to a sustainable society..." and systematically devalues "...both the sanctity of life and the autonomy of the individual" (Butler 2009, with further references). In contrast, *The Limits to Growth, 2025, A Global Forecast for the Next Fourty Years*, and *Reinventing Prosperity* seem to be influenced by the outdated Malthusian Theory of Population. In his *Essay on the Principle of Population*, Malthus had proposed the principle that human populations grew allegedly exponentionally while food production grew at an arithmetic rate (Weiss 2011). Unfortunately, Malthus' views were and still are occasionally misrepresented and abused by misanthropes to propagate targeted population reduction by wars, famine, and the deliberate creation of poor living and working conditions fostering disease and premature death of broad sections of the population (Weiss 2011), although Malthus never proposed such repressive measures but rather promoted universal education for the poor as well as late marriage, abstinence, and fewer children for everyone in favor of a higher standard of living, living himself according to these principles (Parikh 2009). Moreover, Malthus' views must be seen against the background of the time in which he lived (1766–1834), because Malthus could not foresee increases in agricultural production due to technical progress (Weiss 2011).

The 1991 COR report *The First Global Revolution. A Report by the Council of the Club of Rome*, the first report written by the COR itself, originally published by *Pantheon Books* on September 3, 1991, contains some misanthropic and extremist world views:

> ...The collapse of communism in the Eastern European countries and the Soviet Union constitutes a major and unsettling factor... the sudden absence of traditional adversaries has left governments and public opinion with a great void... New enemies, therefore, have to be identified... new weapons devised ...In searching for a new enemy to unite us, we came up with the idea that pollution, the threat of global warming, water shortages, famine... would fit the bill. In their totality and in their interactions these phenomena constitute a common threat which demands the solidarity of all peoples... All these dangers are caused by human intervention, and it is only through changed attitudes and behavior that they can be overcome. The real enemy then, is humanity itself. (King and Schneider 1991)

Unfortunately, such views – epecially if coupled with a call for "population policies" (King and Schneider 1991) – facilitate the occasional classification of climatology as an "ideology" or "religion" (McVeigh 2009). The Pope's interference into the Climate Change topic does not help either. As university-level theologist and trained chemistry technician (Vaticanradio 2013) who is no university-level scientist or engineer, the Pope was called "misguided" by climate scientists for calling Climate Change a "sin" at the World Day of Prayer for Creation in September 2016, stating that human-made Climate Change contributed "...to the heart-rending refugee crisis..." (Bentz 2016) and erroneously "...equating releases of CO2 into the atmosphere as a similar act of pollution as the release of more obvious and truly harmful toxic substances..." (Bentz 2016). The scientists noted that Pope Francis probably got "...terrible advice from some exalted churchmen who are seriously deficient in scientific knowledge" (Bentz 2016). Pope Francis' plan to deliver an encyclical on environmental challenges was criticized by the Catholic US presidential candidate *Rick Santorum*: "The church has gotten it wrong a few times on science...we probably are better off leaving science to the scientists" (Vale 2015).

In a modern democratic society, scientists should be able to do their research and their publications independently without any influence from influential church leaders, politicians, media, industrial lobbyists, or covert political transformers. Especially the Climate Change topic can be easily abused by persons and entities with vested interests. Some consider the Paris Agreement "...a massive transfer of wealth that will have no impact on the climate ...calling attention to warming when the earth is clearly in a cooling phase…" (Corombos 2015). Tom Wigley, a globally respected climatologist, said: "If we introduce the Kyoto Protocol in its original form, that is every country reduced to the amounts that we want, nobody would be able to measure the difference…" (Corombos 2015). According to the climatologist Tim Ball, the true objective of the COP conferences including COP21 had been to establish the *Green Climate Fund*:

> …They decided that developed nations had developed by using fossil fuels and the fossil fuel byproduct, CO2, was causing climate change or initially global warming…Therefore, those 23 developed nations had to pay for their sins…to put money into a fund…to be distributed to the developing nations that were being punished or penalized by those developed nations…At COP 16, they introduced…the Green Climate Fund. That's what they were approving in Paris…a bank account set up in South Korea under the International Monetary Fund, at which all of these developed nations have to put money in to give out to the developing nations. (Corombos 2015)

Therefore, it is important to clarify in how far the Paris Agreement is legally binding (C2ES 2015) or not (Page 2015; Siciliano 2017; Resnick 2017). Already before COP21 in Paris, three countries (Russia, India, and China) had stated that the global warming science was a fraud. Those nations, who are skeptical, are "… poised to economically exploit those 'who are buying into the alleged threat'…" (Corombos 2015).

Nevertheless, such fraud schemes under the cloak of environmentalism should not deter political leaders, business people, and citizens from taking into consideration that fossil energies are indeed a limited resource, in addition to being environmentally rather harmful – at least due to their toxic combustion products like carbon monoxide (CO), nitrous gases (NOx), and sulfur dioxide (SO_2), and eventually also due to the enhanced greenhouse effect by the increased CO_2 levels in the atmosphere which might contribute to Climate Change (Schwarz-Herion 2015a, b, with further references). Therefore, the current trend toward a low-carbon economy has the potential to equally serve the social, ecological, and economic dimensions of SD, if the global civil society seeks the dialogue with the current power players in all fields and demands a higher level of transparency regarding climate science as well as the actual costs and environmental risks of infrastructure projects. In this case, the implementation of the *Sustainable Development Goals* might actually lead to a win-win situation for many stakeholders on a global base.

References

Adams M (2015) The United Nations 2030 Agenda decoded: It's a blueprint for the global enslavement of humanity under the boot of corporate masters. Retrieved from http://www.naturalnews.com/051058_2030_Agenda_United_Nations_global_enslavement.html on June 24, 2017

Beck G, Parke H (2013) Agenda 21. Pocket Books, New York

Bentz J (2016) Scientists say Pope Francis is misguided on 'climate change' as environmental 'sin'. In: Catholic Citizens.org-Catholic citizens of Illinois, October 13,

2016. Retrieved on June 26, 2017, from https://catholiccitizens.org/news/68321/scientists-say-pope-francis-misguided-climate-changeenvironmental-sin/
Bhattacharya A, Ebinger CK, Frank C, Kharas H, Liu W, McArthur J, McKibbin WJ, Meltzer JP, Morris A, Qureshi, Z, Sierra K, Stern N, Sy A, Wilcoxon J (2015) COP 21 at Paris: the issues, the actors, and the road ahead on climate change. Retrieved from https://www.brookings.edu/research/cop-21-at-paris-the-issues-the-actors-and-the-road-ahead-on-climate-change/ on April 28, 2017
Black R (2012) Police end 'Climate Gate' inquiry. On: BBC News. 18 July 2012. Retrieved from http://www.bbc.com/news/science-environment-18885500
Bojanowski A (2010) Heißer Krieg ums Klima. Spiegel online. Published on May 3, 2010. Retrieved from http://www.spiegel.de/wissenschaft/natur/climategate-alles-ueber-den-skandal-in-der-klimaforschung-a-688175.html on June 20, 2017
Bolin B (1970) The carbon cycle. Scientific American, September, p 131
Booker C (2009) Climate change: this is the worst scientific scandal of our generation. The Telegraph online: November 28, 2009. http://www.telegraph.co.uk/comment/columnists/christopherbooker/6679082/Climate-change-this-is-the-worst-scientific-scandal-of-our-generation.html. Accessed 20 June 2017
Boykoff MT, Rajan SR (2007) Signals and noise. Mass-media coverage of climate change in the USA and the UK. EMBO Rep 8(3):207–211. PMCID/PMC1808044-Science and Society. https://doi.org/10.1038/sj.embor.7400924
Broder J (2008) Obama affirms climate change goals. New York Times Online, November 18, 2008. Retrieved on May 21, 2017, from http://www.nytimes.com/2008/11/19/us/politics/19climate.html
Butler S (2009) The dark history of population control. In: Book review: Matthew Connelly: fatal misconception: the struggle to control world population. Harvard University Press, Cambridge, 2008. Retrieved on June 25, 2017, from http://climateandcapitalism.com/2009/11/23/the-dark-past-of-populationconfrom/
Byrne P (2007) Codex Alimentarius to approve 'vitamin guidelines'. Retrieved on June 23, 2017, fromhttp://anhinternational.org/2007/10/10/codex-alimentariusto-approve-vitamin-guidelines/
CBD (2017) History of the convention. Retrieved on May 28, 2017, from https://www.cbd.int/history/default.shtml
C2ES (2015) Outcomes of the U.N. climate change conference in Paris. Retrieved on June 25, 2017, from https://www.c2es.org/international/negotiations/cop21-paris/summary
C2ES (2017) Paris Climate Agreement Q&A. https://www.c2es.org/international/2015-agreement/paris-climate-talks-qa. Accessed 25 June 2017
CIA (Central Intelligence Agency) (2017) Kissinger on foreign policy. Retrieved on June 22, 2017, from https://www.cia.gov/library/readingroom/document/ciardp79b01737a000300010029-0
Coffman M (2009) Explanation of the biodiversity treaty and the Wildlands project. On: Idaho Observer, December 2009. Retrieved on June 12, 2017, from http://www.proliberty.com/observer/20091223.htm
Colombo U (1997) The Club of Rome and sustainable development. http://donellameadows.org/archives/the-club-of-rome-and-sustainable-development/ Accessed 2 May 2017
Cook J, Nuccitelli D, Green S A, Richardson M, Winkler B, Painting R, Way R, Jacobs P, Skuce A (2013). Quantifying the consensus on anthropogenic global warming in the scientific literature. Environ Res Let 8(2). Published on May 15, 2013
COR (2017a) About us. Retrieved on May 26, 2017, from https://www.clubofrome.org/about-us/
COR (2017b) Reports. Retrieved on May 26, 2017, from https://www.clubofrome.org/reports/
Corombos G (2015) Climate expert: Obama's deal a 'scam from top to bottom'. http://www.wnd.com/2015/12/climate-expert-obamas-deal-a-scam-from-top-to-bottom/. Accessed 23 June 2017
Costella J (2010) The Climate Gate E-Mails. The Lavoisier Group Inc. Retrieved from http://www.lavoisier.com.au/index.php on June 20, 2017

Day M (2012) West Germany's CDU had private spy service. West Germany's Christian Democrats created a private spy network staffed with ex-Nazis and aristocrats in a bid to scupper rapprochement with the communist East in the 1970s, a new book claims. In: The Telegraph, December 3, 2012. Retrieved from http://www.telegraph.co.uk/news/worldnews/europe/germany/9719105/West-Germanys-CDU-had-private-spy-service.html. On June 22, 2012
Dr. Rath Foundation (2005) Codex Alimentarius and the GM threat. Retrieved on June 8, 2017, from http://www4.drrathfoundation.org/THE_FOUNDATION/Events/codex-gmthreat.htm
Ehrlich PR, Harte MA, Harwell PH, Raven C, Sagan G, Woodwell M, Berry J, Ayensen ES, Ehrlich AH, Eisner T, Gould SJ, Grover HD, Herrera R, May RM, Mayr E, McKay CP, Mooney HA, Meyers N, Pimentel D, Teal JM (1983) Long-term biological consequences of nuclear war. Science 222:1293–1300
FAO (2017) What is the codex Alimentarius? Retrieved on June 25, 2017, from http://www.fao.org/fao-who-codexalimentarius/en/
Film Atlas (2017) The Noah's ark principle. Retrieved on June 21, 2017, from http://www.atlas-film.com/product/by-genre/classic---cult/the-noahs-ark-principle.html
Fleming J (2017) Operation Gladio. In: Statewatch briefing. Retrieved on June 23, 2017, from http://www.thejohnfleming.com/gladio.html
Gettelman A, Rood RB (2016) Demystifying climate models. Springer Nature, Berlin/Heidelberg
Goñi U (2016) Kissinger hindered US effort to end mass killings in Argentina, according to files. The Guardian. Published on 9 August, 2016. https://www.theguardian.com/world/2016/aug/09/henry-kissinger-mass-killings-argentina-declassified-files
Grefe C (2016) Die Zeit. On 15 Sep 2016. Retrieved from http://www.zeit.de/wirtschaft/2016-09/club-of-rome-nachhaltigkeitsstrategien-globalisierung-jorgen-randers on June 23, 2017
Health Freedom USA (2017) Who is behind codex Alimentarius? Retrieved on June 8, 2017, from http://www.healthfreedomusa.org/aboutcodex/indepth/who.shtml
Hap A (2013) Club of Rome. Opsec News. Exposing the unknown. Retrieved on June 6, 2017, from http://www.opsecnews.com/club-of-rome/
Hickman L, Randerson J (2009) Climate sceptics claim leaked emails are evidence of collusion among scientists. The Guardian online. Amended on 25 Nov 2009. https://www.theguardian.com/environment/2009/nov/20/climate-sceptics-hackers-leaked-emails. Retrieved June 6, 2017
Hoffmann S (1968) Obstinate or obsolete? The fate of the nation-state and the case of Western Europe. In: Hoffmann S (ed) Conditions of world order. Houghton Mifflin, Boston
Hoffmann S, Aron R, O'Connell DP, Gladwyn L, Van Benthem Van Den Bergh G, Kissinger H, Aranguren JLL, Jaguaribe H, Waddington CH, Malek I, Tinbergen J, Fourastié J, Dykman JW, Scrimshaw NS, Gadamer HG, Mus P, Nicol D (1968) Conditions of world order. In: Hoffmann S (ed) Conditions of world order. Houghton Mifflin, Boston
Holbraad M, Petersen MA (2013) Times of security: ethnographies of fear, protest and the future, Routledge studies in anthropology. Tayler and Francis, London
Horn E (2014) Zukunft als Katastrophe. Fischer, Frankfurt am Main
Hower M (2016) Report: so, COP21 happened – what next? Retrieved from https://www.greenbiz.com/article/report-report-so-cop21-happened-what-next on June 8, 2017
Hugendick D (2014) Zukunft als Katastrophe. Wir letzten Menschen. Die Zeit online. Retrieved from http://www.zeit.de/kultur/literatur/2014-07/eva-horn-zukunft-als-katastrophe
IMF (2014) How the IMF promotes global economic stability. Retrieved on June 25, 2017, from https://www.imf.md/imfstabil.html
Inadvertent Climate Modification (1971) Report of the study of man's impact on climate. MIT Press, Cambridge, MA, p 234
IPCC (2017a) IPCC sixth assessment report cycle. http://www.ipcc.ch/. Accessed 22 Apr 2017
IPCC (2017b) History. https://www.ipcc.ch/organization/organization_history.shtml. Accessed 22 Apr 2017
Jaguaribe H (1968) World order, rationality, and socioeconomic development. In: Hoffmann S (ed) Conditions of world order. Houghton Mifflin Company, Boston

Jaguaribe I (2017) É Tudo Verdade: Tudo é Irrelevante. Hélio Jaguaribe. Retrieved on June 25, 2017, from https://carmattos.com/2017/04/24/e-tudo-verdade-tudoe-irrelevante-helio-jaguaribe/
Johnson K (2009) Climate Emails Stoke Debate. Scientists' leaked correspondence illustrates Bitter Feud over Global Warming, 23 Nov 2009. Wallstreet Journal online. Retrieved from https://www.wsj.com/articles/SB125883405294859215 on June 5, 2017
Johnson P (2015) Globalism through U.N.'s Agenda 21, agenda 2030, and vision 2050. https://www.technocracy.news/index.php/2016/08/01/globalism-u-n-s-agenda-21-agenda-2030-vision-2050/
Kaysen C (1972) The computer that printed out W*O*L*F*. Foreign Aff 50(4):660–668. https://doi.org/10.2307/20037939. JSTOR 20037939.
Kellner DM (2009) Cinema wars: Hollywood film and politics in the Bush-Cheney era. Wiley, Chichester
King A, Schneider B (1991) The First Global Revolution. A Report by the Council of the Club of Rome. Pantheon Books, New York City
King A, Okita S, Peccei A, Pestel E, Thiemann H, Wilson C (1972) Commentary to the limits to growth. Universe Books, New York
Kissinger HA (1968) Domestic structure and foreign policy. In: Hoffmann S (ed) Conditions of world order. Houghton Mifflin, Boston
Lexikon der Nachhaltigkeit (2015) Weltgipfel Rio. Retrieved from https://www.nachhaltigkeit.info/artikel/rio_weltgipfel_1437.htm on April 28, 2017
Machta L (1971) The role of the oceans and biosphere in the Carbon Dioxide Cycle. Presented at Nobel symposium 20 "The Changing Chemistry of the Oceans", Gothenburg, Sweden, August 1971
Maguire J (2008) Roland Emmerich & 10,000 B.C. http://maguiresmovies.blogspot.de/2008/04/roland-emmerich-10000-bc.html. Accessed on 7 June 2017
Masters (2017) Al Gore's an inconvenient truth. Retrieved on June 14, 2017, from https://www.wunderground.com/resources/education/gore.asp
Matsangou E (2017) The politics of the IMF. The IMF frequently steps in to offer financial aid when countries need it most, but how often does politics play a part in the solutions it supplies? On: world finance-the voice of the market, April 10, 2017. Retrieved on June 25, 2017, from https://www.worldfinance.com/infrastructure-investment/the-politics-behind-the-imf
Maxeiner D, Miersch M (2011) Bis 2050 wird der "Normalbürger" abgeschafft. In: Die Welt online, September 1, 2011. Retrieved on June 23, 2017, from https://www.welt.de/debatte/kolumnen/Maxeiner-und-Miersch/article13578319/Bis-2050-wird-der-Normalbuerger-abgeschafft.html
McVeigh K (2009) Judge rules activist's beliefs on climate change akin to religion. The Guardian online. 3 November 2009. https://www.theguardian.com/environment/2009/nov/03/tim-nicholson-climate-change-belief
Meadows DH, Meadows DL, Randers J, Behrens WW (1972) The limits to growth. Universe Books, New York
Mittelbach A (2013) Die sieben Säulen der Freiheit – Vom modernen Sklaven zum unabhängigen Menschen. CreateSpace Independent Publishing Platform, North Charleston
Monbiot G (2009) Pretending the climate email leak isn't a crisis won't make it go away. On: the guardian online. November 25, 2009. Retrieved on June 6, 2017, from https://www.theguardian.com/environment/georgemonbiot/2009/nov/25/monbiot-climate-leak-crisis-response
Moore N, Terbeek N, Thym N (1999) International financial crisis. Retrieved on June 25, 2017, from https://web.stanford.edu/class/e297c/trade_environment/global/hcrisis.html
NASA (2017) Scientific consensus: Earth's climate is warming. Retrieved from https://climate.nasa.gov/scientific-consensus/ on June 21, 2017
NSA Archive (2012) Sunshine Week Document Friday! Kissinger says, "The illegal we do immediately; the unconstitutional takes a little longer. But since the FOIA, I'm afraid to say things like that." Retrieved from https://nsarchive.wordpress.com/2012/03/15/document-friday-kissinger-says-the-illegal-we-do-immediately-the-unconstitutional-takes-a-little-longer-but-since-the-foia-im-afraid-to-say-things-like-that/ on June 26, 2017

Page S (2015) No, The Paris Climate Agreement isn't binding. Here's why that doesn't matter. Climate Reporter. Published on 14 Dec 2015. https://thinkprogress.org/no-the-paris-climate-agreement-isnt-binding-here-s-why-that-doesn-t-matter-62827c72bb04

Parikh C (2009) Thomas Malthus. In: Capitalideas Online, November 25, 2009. Retrieved from https://capitalideasonline.com/wordpress/thomasmalthus/. On June 24, 2017

Perkins J (2005) Confessions of an economic hit man. Ebury Press/Random House, London

Platt J (2013) The Great Sparrow Campaign was the start of the greatest mass starvation in history. In 1958, Mao Zedong ordered all sparrows to be killed. As a direct result, millions of people starved to death. On: Mother Nature Network. September 30, 2013. Retrieved from https://www.mnn.com/earth-matters/animals/stories/the-great-sparrow-campaign-was-the-start-of-the-greatest-mass on June 21, 2017

PPO (2017a) 31,487 American scientists have signed this petition, including 9029 with PhDs. Retrieved from http://www.petitionproject.org/ on May 4, 2017

PPO (2017b) Qualifications of signers. http://www.petitionproject.org/qualifications_of_signers.php. Accessed 4 May 2017

PPO (2017c) Summary of peer-reviewed research. Retrieved from http://www.petitionproject.org/review_article.php on May 5, 2017

PPO (2017d) Frequently asked questions. http://www.petitionproject.org/frequently_asked_questions.php. Accessed 4 May 2017

Pusa M (2017) Mao Zedong's Sparrow Campaign caused one of the worst environmental disasters in history. On: Elite Readers. Retrieved from https://www.elitereaders.com/mao-zedong-great-sparrow-campaign-worst-environmental-disaster-history-china-famine/ on June 8, 2016

Raman M (2016) The Climate Change battle in Paris: an initial analysis of the Paris COP21 and the Paris Agreement. Third World Network on February 2, 2016. https://twn.my/title2/climate/info.service/2016/cc160202.htm. Accessed 4 June 2017

Randers J (2012) 2052 a global forecast for the next forty years. Chelsea Green Publishing, Vermont

Randers J, Maxton G (2016) Reinventing prosperity: managing economic growth to reduce unemployment, inequality, and climate change. Greystone Books, Vancouver

Ravkin A (2009) Hacked E-Mail is new fodder for climate dispute. Retrieved from http://www.nytimes.com/2009/11/21/science/earth/21climate.html?mcubz=1 on June 6, 2017

Resnick B (2017) 4 things to know about the Paris climate agreement. Retrieved on June 24, 2017, from https://www.vox.com/energy-andenvironment/2017/6/1/15724162/trump-paris-climate-agreement-explained-briefly

Ribeiro S (2016) The fourth industrial revolution, technologies and impacts. Retrieved from http://www.alainet.org/en/articulo/181503

Ritchie EJ (2016) Fact checking the claim of 97% consensus on Anthropogenic Climate Change. Forbes online 2016. Published on December 14, 2016. Retrieved from https://www.forbes.com/sites/uhenergy/2016/12/14/fact-checking-the-97-consensus-on-anthropogenic-climate-change/#3996e4e61157 on May 5, 2017

Rivera D (1994) Modern history project. The origin of the Club of Rome. Retrieved from http://modernhistoryproject.org/mhp?Article=FinalWarning on April 30, 2017

Schwartz P, Randell D (2003) An abrupt Climate Change scenario and its implications for United States Security. http://eesc.columbia.edu/courses/v1003/readings/Pentagon.pdf. Accessed 16 June 2017

Schwarz-Herion O (2005) Die Integration des Nachhaltigkeitsgedankens in die Unternehmenskultur und dessen Umsetzung in die betriebliche Praxis. Eine empirische Studie zur ökologischen und sozialen Verantwortung von Privatunternehmen. Shaker, Aachen

Schwarz-Herion O (2015a) From conventional worlds to a New Sustainability Paradigm (NSP): Raskin's model scenarios in the light of current trends. In: Schwarz-Herion O, Omran A (eds) Towards the new sustainability paradigm. Managing the great transition to global democracy. Springer, Cham

Schwarz-Herion O (2015b) Urgent ecological problems for the NSP. In: Schwarz-Herion O, Omran A (eds) Towards the new sustainability paradigm. Managing the great transition to global democracy. Springer, Cham

Shabecoff P (1988) Global Warming has begun, Expert tells senate. The New York Times, June 24, 1988
Shubik M (1972) Modeling on a grand scale. Science 74(1971):1014ff
Siciliano J (2017) Trump on Paris Climate Agreement: 'Like hell it's non-binding'. Washington Examiner. Published on June 21, 2017. http://www.washingtonexaminer.com/trump-on-paris-climate-agreement-like-hell-its-non-binding/article/2626729
Snyder M (2015) The 2030 agenda: this month the UN launches a blueprint for a new world order with the help of the Pope. http://endoftheamericandream.com/archives/the-2030-agenda-this-month-the-un-launches-a-blueprint-for-a-new-world-order-with-the-help-of-the-pope
Solow RM (1973) Is the end of the world at hand? Challenge 16(1):39–50. https://doi.org/10.2307/40719094.JSTOR 40719094
Soltan L (2016a) Sad facts on technotrash or e-waste. http://www.digitalresponsibility.org/sad-facts-on-technotrash. Accessed 6 June 2017
Soltan L (2016b) Environmental and societal impact of technology. http://www.digitalresponsibility.org/environmental-and-societal-impact-of-technology/. Accessed 6 June 2017
Spencer R (2007) An inconvenient truth: blurring the lines between science and science fiction. GeoJournal 70:11–14. https://doi.org/10.1007/s1007/s10708-008-9129-9
Sputnik (2017) The uncertainty has settled. Retrieved from https://www.sputnik-kino.com/program/events on June 26, 2017
Staud T (2015) Zeitleiste: Die internationalen Klimaverhandlungen – eine Chronik. http://www.bpb.de/gesellschaft/umwelt/klimawandel/200832/zeitleiste-die-internationalen-klimaverhandlungen-eine-chronik. Accessed 24 April 2017
Stern N (2007) The stern review. Cambridge University Press, Cambridge
Stern N (2015) Why are we waiting? The logic, urgency, and promise of tackling climate change. MIT Press Cambridge, Cambridge, MA
Suter K (1999) Fair warning. The Club of Rome revisited. the story. Accessed 3 May 2017. http://www.abc.net.au/science/slab/rome/
The Guardian (2010) Climate emails: were they really hacked or just sitting in cyberspace? Slack security or subversion at the university may have led to 'unintentional sharing', making the police investigation pointless. https://www.theguardian.com/environment/2010/feb/04/climate-change-email-hacker-police-investigation. The Guardian online. February 4, 2010. Accessed 7 June 2017
The Guardian (2017) Spoiler alerts: the five best climate-change films. Retrieved from https://www.theguardian.com/environment/2017/jan/18/best-top-climate-change-films. On June 23, 2017
The Observer (2001) Saudi dove in the oil slick. Published on January 14, 2001. https://www.theguardian.com/business/2001/jan/14/globalrecession.oilandpetrol.. Accessed 7 May 2017
Thompson S (1982) The Anglo-Soviet circles that created Kissinger. In: EIR, Vol. 9, N° 36, September 21, 1982. Retrieved from https://de.scribd.com/document/326317333/Anglo-Soviet-Circles-That-Created-Kissinger-by-Scott-Thompson-6. On June 21, 2017
Tinbergen J (1968) International economic planning. In: Hoffmann S (ed) Conditions of world order. Houghton Mifflin, Boston
Tinbergen J, Dolman AJ, Ettinger JV, Club of Rome (1976) Reshaping the international order: a report to the club of Rome. Dutton, Boston
Todhunter C (2017) Codex Alimentarius and Monsanto's toxic relations. Retrieved on June 25, 2017, from http://www.countercurrents.org/2017/07/21/codexalumentarius- and-monsantos-toxic-relations/
Trump D (2017) Statement by president trump on the Paris climate accord. In: Garden (ed) Office of the press secretary immediate release. The White House, Washington, DC
Turco RP, Toon OB, Ackerman TP, Pollack JB, Sagan C (1983) Nuclear winter: global consequences of multiple nuclear explosions. Science 222:1283–1292
Tuttle I (2015) The 97 percent solution. National Review online. Published on October 8, 2015. Retrieved from http://www.nationalreview.com/article/425232/97-percent-solution-ian-tuttle on April 30, 2017

UN (2017a) Goal 13: take urgent action to combat climate change and its impacts. Retrieved from http://www.un.org/sustainabledevelopment/climate-change-2/ on April 22, 2017
UN (2017b) Open Working Group proposal for Sustainable Development Goals resolution and report. A/68/L.61 – GA takes action – report of the Open Working Group on Sustainable Development Goals. Retrieved from https://sustainabledevelopment.un.org/focussdgs.html on April 23, 2017
UN (2017c) Agenda 21 UNCED, 1992. Retrieved from https://sustainabledevelopment.un.org/outcomedocuments/agenda21 on April 30, 2017
UN (2017d) Special session of the general assembly to review and appraise the implementation of agenda 21, 23–27 June 1997, New York. http://www.un.org/esa/earthsummit/. Accessed 23 Apr 2017
UN (2017e) Earth summit + 5 special session of the UN General Assembly, 23–27 June 1997, New York. http://www.un.org/en/events/pastevents/earthsummit_plus_5.shtml. Accessed 24 Apr 2017
UN (2017f) Sustainable development-Cop 21. Retrieved from http://www.un.org/sustainabledevelopment/cop21/
UN (2017g) Transforming our world: the 2030 Agenda for Sustainable Development. Retrieved from https://sustainabledevelopment.un.org/post2015/transformingourworld on June 23, 2017
UNFCCC (2017a) Paris Agreement – status of ratification Retrieved from http://unfccc.int/paris_agreement/items/9444.php on April 25, 2017
UNFCCC (2017b) Marrakech climate change conference – November 2016. http://unfccc.int/meetings/marrakech_nov_2016/meeting/9567.php. Accessed 25 Apr 2017
UNFCCC (2017c) The Second World Climate conference. https://unfccc.int/resource/ccsites/senegal/fact/fs221.htm. Accessed 25 Apr 2017
UNFCCC (2017d). Conference of the Parties (COP2). Retrieved from http://unfccc.int/bodies/body/6383.php on April 24
United Nations Department of Economic and Social Affairs (1970). Statistical yearbook 1969: New York
Vale P (2015) Pope Francis told to stay out of climate change by devout catholic presidential candidate rick santorum. Huffington Post Online 03/06/2015 16:24. Updated 3 June 2015. http://www.huffingtonpost.co.uk/2015/06/03/pope-francis-told-to-stay-out-of-climate-change-_n_7502772.html
Van Benthem Van Den Bergh G (1968) Contemporary nationalism in the western world. In: Hoffmann S (ed) Conditions of world order. Houghton Mifflin, Boston
Van der Reijden J (2013) The supranational suspects behind 9/11: white house, CIA, Suadis, Pakistanis, a Russian GRU Firm, and The. In: Institute for the study of globalization and covert politics-9/11 study center. Retrieved on June 23, 2017, from https://isgp-studies.com/911-supranational-suspects
Vaticanradio (2013) Kurzbiographie des neuen Papstes: Papst Franziskus. http://de.radiovaticana.va/storico/2013/03/13/kurzbiographie_des_neuen_papstes_papst_franziskus_/ted-673086
Weingart P, Engels A, Pansegrau P (2000) Risks of communication: discourses on climate change in science, politics, and the mass media. Public Underst Sci 9(3):261–283
Weiss L (2011) Die Malthusische Bevölkerungstheorie - Malthusian Population Theory. Retrieved from https://vorleungen.info/node/1218. On June 24, 2017
Wikipedia (2017) Media coverage of climate change. Retrieved on May 5, 2017, from https://en.wikipedia.org/wiki/Media_coverage_of_climate_change
Wiki-Talk (2017) Talk. Media coverage of climate change. Retrieved on May 5, 2017, from https://en.wikipedia.org/wiki/Talk:Media_coverage_of_climate_change
Yerman MG (2016) Climate change and the media. Huffington Post. Published on July 29, 2016. Retrieved from http://www.huffingtonpost.com/marcia-g-yerman/climate-change-and-the-me_1_b_11207298.html

Chapter 2
Factors Contributing to the Catastrophic Flood in Malaysia

Abdelnaser Omran, Odile Schwarz-Herion, and Arpah Abu Bakar

Abstract In 2014, a flood hit a Kelantan state located in the East Coast part of Malaysia, causing a massive destruction which left this state paralyzed. The National Security Council (NSC) confirmed that the massive flood that hit Kelantan was the worst in the history of the state. Most of the divisions of that state were flooded with water with high content of mud. Thus, this study was conducted to look at the factors that cause a catastrophic flood in Kelantan state, and it was particularly focused on one of the districts in the State, named Kuala Krai district. In conclusion, it was found that equally natural and human factors contributed the catastrophic flood in Kelantan state. This study led to several useful recommendations; one of these is to provide an awareness program for the residents in Malaysia on such disasters. This program can provide sufficient knowledge and awareness to the residents about the important steps or actions that can be taken in case of a flood in order to stay safe. The potential impact of weather modification technology, especially cloud seeding, on the Kelantan flood is shortly addressed, although further details of this risky technology would go beyond the scope of this chapter.

Keywords Natural factors · Human factors · Cloud seeding · Catastrophic · Kelantan flood 2014 · Malaysia

2.1 Introduction

Flood is defined as a temporary inundation of normally dry land areas from the overflow of inland or tidal waters or from the unusual and rapid accumulation or runoff of surface waters from any source. The rise in water may be caused by excessive

A. Omran (✉) · A. Abu Bakar
School of Economics, Finance & Banking, Universiti Utara Malaysia, Sintok, Kedah, Malaysia
e-mail: naser_elamroni@yahoo.co.uk

O. Schwarz-Herion
Dres. Shwarz-Herion KG Detective Agency, Ettlingen, Germany

A. Omran, O. Schwarz-Herion (eds.), *The Impact of Climate Change on Our Life*, https://doi.org/10.1007/978-981-10-7748-7_2

rainfall, snowmelt, natural stream blockages, and windstorm over a lake or any combination of such conditions (Pure Treatment Plant 2017). In science, flood is defined as a temporary rise of the water level, as in a river or lake or along a seacoast, resulting in its spilling over and out of its natural or artificial confines onto land that is normally dry. Floods are usually caused by excessive runoff from precipitation or snowmelt or by coastal storm surges or other tidal phenomena (Dictionary 2015). In general, flood is defined as a temporary overflow of a normal dry area due to overflow of a body of water, unusual build up, runoff of surface waters, or abnormal erosion or undermining of shoreline. Floods can also be an overflow of mudflow caused by a buildup of water underground (Dictionary 2015). Over the years, flood has become one of the most destructive disasters in Malaysia. Normally, every year, Malaysia faces flood during the northeast monsoon season, and it is popular in the East Coast state of Peninsular Malaysia which is Kelantan, Pahang, and Terengganu.

In our study, we will focus on the catastrophic flood that hit Kelantan at the end of year 2014. Talking about flood in Kelantan, there are many catastrophic flood histories that happened in Kelantan. One of them is the disaster of Bah Merah. The Bah Merah is a legend in the past and being delivered generation by generation. Bah Merah happened in year 1926 caused by heavy rains for 10 days without stopping, and these heavy rains led to a catastrophic flood in Kelantan during that year. Why it is called or known as Bah Merah? According to history, it is because of the flood water mixed with headwaters collapse and the water change to red color like tea. Due to that reason, the residents called it as Bah Merah disaster. About 537,000 persons or 84% of the residents were affected. After that, another catastrophic flood happened again in year 1967 and also in the year 2004. The latest flood incident happened in the previous year, where the flood which happened was similar to the Bah Merah disaster which damaged almost the whole state of Kelantan. Flood in Kelantan in year 2014 was the worst flood happened in Peninsular Malaysia. This flood was classified as annual flood because of its frequency where it happens in every end of the year during the northeast monsoon season, but by comparing the magnitude of all the floods that happened in the past, this incident was the worst one. The flood was caused by heavy rainfall(s) brought about by the northeast monsoon to the Peninsular Malaysia. The entire district in Kelantan was affected by floods except in Bachok. The flood affected the district of Tumpat including Jeli. The worst affected area was the district of Gua Musang and Kuala Krai. The flood that happened last year had claimed many lives and destroyed a lot of property. This is really a big loss to the citizen who suffered from this disaster. Statistics shows that the total number of flood victims transferred to the evacuation centers in states involved with the flood disaster increased during that time. Overall the total victims were 242,046 persons involving five states, and the data were collected until 7 am on 30 December 2014. The total victims transferred to the evacuation centers in Kelantan with 158,476 peoples, Pahang with 49,978 peoples, Terengganu with 25,775 peoples, Perak with 7664 people, and Johor the least with 153 persons, respectively (Malaysia Mail Online 2014). We can see that the total victims transferred to the evacuation centers in Kelantan had the highest total of flood victims compared to the total flood victims to the evacuation centers in Pahang, Terengganu, Perak, and Johor during that time. The percentage victim in Kelantan was 65% from the overall total of flood victims. In this case, the total showed us about how worst the flood in Kelantan was. We strongly believe that the actual total number of flood victims might exceed the number of victims registered to the local authorities. Based on reports by the various sources, there are many reasons that led to the flood in Kelantan. The reasons are categorized into two main factors that are the natural factor and the human factor. According to reports, the natural causes resulted from the monsoon season that is the northeast monsoon and climate changes, whereas the human factors include negligent human factors such as land cleaning, waterways, unmanaged drainage, development, and possibly also deliberate human factors like the use of weather modification technology in the form of cloud seeding (Vidal and Weinstein 2001) which is only shortly addressed in this chapter and will be explained and discussed in detail in a later chapter toward the end of this book.

2.2 Problem Statement

Malaysia is a country free from other disasters such as volcanic eruption and earthquake, but flood is a common natural disaster that the country is always facing. Floods in Malaysia are severe since the country lies on the equator range. As a country with a monsoon season, Malaysia receives high amount of rainfalls from November to March every year when it is the northeast monsoon. Flood in Malaysia occurs annually where it becomes the concern of the government. At the end of year 2014, Malaysia was hit badly by the catastrophic flood – a disaster which was more severe than the flood in 1971. The flood in year 1971 led to massive property destructions, losing RM200,000 million, and left 61 people dead (The Star 2015). This amount increased in the 2014 flood where property damages and infrastructure rose up to one billion ringgits with 21 deaths (The Star 2015). From this, Kelantan was the most affected state in Malaysia. The National Security Council (NSC) confirmed the massive flood that hit Kelantan was the worst in the history of the state. The flood caused a massive destruction in Kelantan which had that state paralyzed. Most of the divisions of that state were flooded with water with high content of mud. In order to solve this problem, it is the responsibility of every single party to take their own actions on preventions and precaution(s). With the combined actions from every party, flood can be reduced or avoided from taking place in Malaysia, especially Kelantan itself. This research is intended to find out the factors that cause flood in Malaysia. The factors that are focused in this research are divided into two categories: natural factors and human factors (because these two categories are equally relevant to the causes of flood).

The human factors can be subdivided into the negligent and the deliberate human factors. While most current reports on the Kelantan flood seem to be limited to the natural factors and to the negligent human factors (land cleaning, waterways, unmanaged drainage, and development), the catastrophic flood from 1971 in Malaysia and the Kelantan flood from 2014 in this magnitude might very well have been caused by the deliberate human factor: human-made weather modification which was already applied in field experiments since at least the 1940s (Rich 2011).This kind of technology can be divided into three major categories: (1) suppression of weather patterns, (2) intensification of weather patterns, and (3) the introduction of completely new weather patterns; the two basic methods of weather modification include cloud seeding and directed energy (Schwarz-Herion 2015, with reference to The Royal Society 2009; Rich 2011).

2.3 Past Studies

2.3.1 Flood

Flood is a natural disaster which happens in many countries. It can be defined as a condition where two or more acres of place are inundated by the water or mudflow (FEMA 2015). Most of the countries in the world suffer from flood with Bangladesh as the country that experienced most flooding. This is due to the condition of that country itself which is located in the low area with over 230 rivers. Effects of flood may be severe depending on the capacity of the flood. In April 1993, Mississippi, USA, and its 50 tributaries have been swelled by the flood due to the excessive rainfall which was ten times more than usual (Natural Disaster Association). This disaster led to huge losses for that country. Countries that lie near the river floodplains and low-lying coastal areas are usually the ones that will suffer flood disaster. In Malaysia, flood occurs annually; every year there will be flood, usually during the monsoon season which is from November to March. The history of floods in Malaysia has started in 1971 (Khalid and Shafiai 2015). Floods have been categorized as the most severe natural disaster in Malaysia since Malaysia is a country that is free from other disasters such as typhoon and volcanic eruption. Supposedly, 90% of the damages associated with the natural disaster in Malaysia are caused by floods in terms of human and economic loss (Ali Khan et al. 2004). Floods in Malaysia can be divided into two types: (1) flash flood and (2) monsoon flood. Flash flood usually takes place in the cities in which there are rapid developments of buildings such as Kuala Lumpur. It happens frequently when there are rains for a few hours. The monsoon flood happens due to the flows of the northeast monsoon which brings heavy rainfall especially to the East Coast states of Peninsular Malaysia and north and southern part of Sabah Sarawak. Based on the Malaysian Department of Irrigation and Drainage, 29,800 km^2 of lands are exposed to the flood risk and 42.2 million people are affected by it (Department of Irrigation and Drainage Malaysia 2009). In Malaysia, there are a total of 189 river basins that direct toward South China Sea with 85 of them exposed to the repeated flood (D/iya et al. 2014). At the end of 2014, Malaysia was hit by the most severe floods in 30 years which had a huge impact on several states and displaced over 200,000 people (Disaster hits Malaysian shores 2015). The East Coast states that are usually affected by the flood in Malaysia are Kelantan, Pahang, and Terengganu with Kelantan. As for our area of study, Kelantan was located in the latitude of 06°10°N and longitude of 102°20′E (Ali

Khan et al. 2014). The position of Kelantan which is near to the South China Sea and the geographical characteristics plus the unplanned urbanization had increased its exposure to the monsoon flood. The 2014 flood was recorded as the worst flood in Kelantan, (even) worse than the big flood of 1967 (Indaramalar and Soon 2015). It is called "Bah Kuning" because the flood was yellow in color due to the excessive mud contents (Baharuddin et al. 2015). During that disaster, almost 1827 homes were washed away, and many schools and premises were destroyed. This had a big impact to the residents – especially to the poor who lived in wooden houses, suffering a total loss as their houses were totally inundated due to the flood. For those houses made of bricks, their houses might not be lost totally, but there will certainly be some trails of the flood even though they are not too severe. In Manek Urai town, 138 out 502 households have suffered a total loss, whereas another 271 only suffered certain damages that only needed some repair (Indaramalar and Soon 2015). Approximately, 1827 houses in Kelantan were destroyed by that flood. Based on the report by the Council, during that severe flood the water level of Sungai Kelantan which can usually only load up to 25 m reached the level of 34.17 m, so that it can be classified as the red level compared to the 33.61 m in the last severe flood in 1967 (Azlee 2015). This disaster reportedly displaced more than 100,000 people in that state (Malaysiakini 2014). It is equated with the tsunami disaster where 202,000 victims were displaced (Baharuddin et al. 2015). Record shows that the rainfall of 1295 mm is usually equivalent with the amount of rain that can be seen for the period of 64 days (Baharuddin et al. 2015). This excessive amount of water increased the water level from three main rivers in Kelantan: Sungai Galas, Dabong; Sungai Lebir, Tualang; and Sungai Kelantan with the record of 46.47 m, 42.17 m, and 34.17 m, respectively. The increase in these three rivers had placed Kelantan in the dangerous situation which in the end led to that disaster. The question that comes across in our minds is, why all this happened? Here is where theories of flood causes arise. Every country in the world will usually try as much as they can in order to avoid the flood disaster which will have a bad impact on the economy and other spheres of life. The aftermath effects of flood are severe. The country did not only experience losses in term of lives of its citizen and properties but also in terms of monetary loss due to the necessity to replace or repair those losses. This problem becomes the concern of Malaysia itself as the country whose wet and dry season increases the probability of floods to occur; this goes especially for Kelantan.

2.3.2 Natural Factors

Based on the business dictionary, natural environment refers to the climate, weather, and natural resources because these three elements contribute or have an impact on the human continuity and economic activities (natural environment). Nature is something that cannot be predicted by any single person. It is under the power of Almighty, but human interference – involuntary by activities like land cleaning or voluntary activities like economic warfare by climate engineering or even targeted ecological warfare by hostile use of weather modification tools – can increase natural disasters significantly. As we know, different countries have different natural environments. In Korea, for example, they experience four different seasons in 1 year, i.e., summer, winter, spring, and autumn (Weather), whereas in Uganda there are only two seasons, that is, the wet season and the dry season (Permin 1997). So, we can say that natural factors are different from one country to another. In Malaysia, there are two seasons, that is, the wet season and the dry season; the wet season lasts from November to March with November being the wettest month, and the dry season lasts from May to September with June as the driest month (World Weather Online 2012). Those mentioned seasons are also known as monsoons: the dry season is called "southwest monsoon" and the wet season is called "northeast monsoon." The transition period between these two monsoons is called "inter-monsoon period" (Ooi et al. 2011; Tan et al. 2015). The northeast monsoon will bring heavy rainfall usually affecting the Peninsular Malaysia more than Western Malaysia which is Sabah and Sarawak. This is because Peninsular Malaysia lies in the equatorial zone where here the climate is hot and wet all year round. Monsoon occurs when there is a difference between land and sea temperature caused by the sun's radiation or climate change. During the winter, the continental landmass will be cooled, resulting in low temperature, and when this happens, the atmospheric pressure will be increased, and an intense high pressure will be created over Siberia. North westerlies (cold air of Siberia) will flow toward Southeast Asia, and it will be changed into north easterlies when reaching coastal water in China. When the cold water interacts with the low pressure, atmospheric systems and cyclonic vortices rapidly will produce strong winds and high seas in South China and heavy rainfall to East Coast states of Peninsular Malaysia, also in Sabah and Sarawak (Ooi et al. 2011). The East Coast states that are usually affected by this monsoon are Kelantan, East Johor, Terengganu, and Pahang, resulting in a severe flood. We choose Kelantan as the case site investigation because from all these stages, Kelantan is most affected by the monsoon. Almost every year, there is a flood in Kelantan, and it starts to become known as common annual floods. This disaster is caused by the heavy rainfall from the monsoon (Ali Khan et al. 2004). Besides this heavy rainfall of the monsoon, another problem arises when there is rain for a long period which is normal during monsoon season and which will decrease the ability of the soil to absorb the water which could also lead to a flood. The types of soil itself have their own roles. If the soil is compacted, then it will be difficult to absorb the water.

2.3.3 Human Factors

Human factor, also known as ergonomic, is associated with the interaction between humans and other elements of the system (HFES 2015). In an easy word, human factors or actions are related to other things in systems. For example, when human does combustion, it will contribute to the greenhouse effect, so as it might have happened with the flood, judging by its frequency and its increased severity over the years due to environmental degradation like deforestation and exploitation of land (Vincent 1997). Nowadays, urbanization has taken place in every country. As the development in technology is increasing rapidly, the same is happening with the development of every country. More forest is exploited to build buildings and houses. This situation will increase the impermeable surfaces. The impermeable surfaces cannot absorb water during rainfall. Humans are also involved in deforestation thus reducing the amount of water intercepted and increasing the run off which could lead to landslides. Human intervention in ecology will have a bad impact as the equilibration of ecology is interrupted.

2.4 Research Method

This research focused on one of the districts in Kelantan state, named Kuala Krai district. An interview session was conducted on 26 September 2015 with 15 respondents including the President of Security Alliance Village Development Committees of Kampung Biak and the villagers who experienced the flood last year. The villagers were selected from one of the villages which were impacted by the flood in the district of Kuala Krai. The interviews were semi-structured and conducted in person. Within the interview sessions, there were two parts consisting of 16 questions each and covering all aspects related to the research issue. The first part of the interview questions was on the general view of evacuees about the flood, and it contained 11 questions; these questions were designed in a way that concentrated on the opinion of residents about the factors which had led to the flood, the party who was responsible during the flood, the management of security, food supply, evacuation centers, and the aids that they had received after the flood. The cost of losses due to the flood and the time taken to recover from the flood have also been asked during the interview sessions. Additionally, the researchers also asked the residents about their experience during the flood – what they have been through before and how they faced or handled it. The second part of the interview questions contained more in-depth questions focusing on the residents' properties. It included about five questions, and a study by Morris-Oswald and Simonovic (1997) helped the researchers in the abovementioned five questions which were on the assessment of the social impact of flooding. Some of the questions referred to the time when the property was built, the period of time for which respondents had lived in the property, the proximity of the property to the nearest river, and any insurance that they had bought for the property. In this part, the focus was more on the loss of property of residents involved in floods and possible ways in which they protect their property or seek compensation for their losses. In addition to that, site inspection visits were also made to the involved places on 27 September 2015 where the researchers visited the places that were affected by the flood. This was done in order to identify the risk that the place was exposed to. For example, the researchers visited some houses affected by the floods just to make "real" discoveries of potential loss exposures of the places and to see whether requirements of safety were met so that prompt actions can be highlighted. In addition, the researchers had some photos made as evidence at the scene.

2.5 Analysis of the Results

Based on various sources of reports and studies, our findings regarding the causes of the catastrophic flood that happened in Kuala Krai, Kelantan, at the end year of 2014 can be classified into two categories: (1) natural causes and (2) human factors that contribute to the incident. These causes are explained in deep details below:

2.5.1 Natural Factors

2.5.1.1 Northeast Monsoon Season

Malaysia is located in the geographical area where it is part of the northeast monsoon. The northeast monsoon happens due to the northeast wind flow from the South China Sea bringing along an amount of moisture in the air. Usually, the northeast monsoon brings along heavy rain that lasts many days (NASA). This, however, increases the risk exposure to flood happening in Malaysia every year. Due to the moisture condition and heavy rain brought about by the northeast monsoon, Kuala Krai, Kelantan, was one of the worst hit states due to the catastrophic flood that hit Malaysia at the end of 2014. According to the Ministry of Science, Technology and Innovation, the northeast monsoon weather happens every year beginning at the month of November until the month of March. The northeast monsoon brings heavy rainfall during the end of the year – especially to the states which are located at the East Coast states in the Peninsular Malaysia (Ministry of Science). Due to the northeast monsoon that strikes Malaysia every year, the research area is located in the area that is vulnerable and prone to flood risk. In addition to that, the unnatural phenomena of excessive rainfall that flooded the Kuala Krai area and included other areas in Malaysia also resulted from the Madden-Julian oscillation (MJO). As stated by meteorology expert, Dr. Liew Ju Neng, Madden-Julian oscillation (MJO) is a tropical climate element that originates from the Indian Ocean and passes through Malaysia four times a year. In conjunction to that, the combination of this element with the northeast monsoon has resulted in an excessive pattern of rainfall due to the east winds that carry more moisture than usual (Mahavera 2015).

2.5.1.2 Climate Changes

Climate change is one of the worrying factors nowadays. The climate change is usually defined as long-term changes toward the climate. Climate changes directly contribute to the extreme changes of weather and temperature. Climate changes cause greater temperature changes that lead to greater storms, stronger monsoons, and heavier rains. This indirectly contributes to extreme weather events, increased sea levels, and the melting of glaciers, all of which can lead to an increase in the number of flooding disasters as it happened in Kelantan in 2014 (Crabtree 2012; Wan Ahmad and Abdurahman 2015). Climate changes result from various human factors that eventually give negative impacts to the environment. In the past few decades, our country has experienced climate change through increasing temperature, changing rainfall patterns, and extreme weather conditions that lead to an increasing amount of climate-related disasters (Sakar et al. 2013). This flood incident in Kelantan is consistent with the global warming trend that has been a huge issue. Human factors, however, contributed to the climate changes. When human does combustion, for example, it contributes to the greenhouse effect. The same impact goes to the flood incidents where the frequency and severity are increasing over the years due to the environmental degradation resulted by activities from deforestation and exploitation of land (Vincent 1997).

2.5.2 Human Factors

2.5.2.1 Land Clearings

Land clearing is one of the main factors that contributed to the massive flood that hit Kuala Krai, Kelantan, at the end of 2014. Based on a speech by Datuk Seri Shahidan Kassim, a minister in the Prime Minister's Department, the land clearing has given a huge contribution in the flooding event in Kuala Krai, Kelantan (Bernama 2015). Furthermore, the Gua Musang MP, Tengku Razaleigh Hamzah, also added to the press by stating that the illegal land clearing was possibly one of the main reasons that contributed to the flood by making the water level rising faster than expected (Bernama 2015). Land clearing resulting from the uncontrollable logging and clearing for agricultural purposes by irresponsible parties has left the state of Kelantan in a condition where it is unable to handle the amount of heavy rainfall at the end of every year. Land clearing impacts the land around, causing flood due to heavy rain. Normally, trees that grow in the land areas act as a natural retention to the excessive rainfalls which is significant during the northeast monsoon season. Unfortunately, land clearing has decreased the amount of trees that usually also act as a natural support to the soil. The decreasing amount of trees in the area of Kuala Krai that is known as an agricultural area for palm oil has increased the risk of flood. The loss of trees resulting from massive land clearing has given a negative impact on the soil structure (Butler 2012). Heavy rain in the monsoon season increases the risk of flood because rainfalls on cleared forest land carry rocks and soil into river, thus causing river sedimentation (Butler 2012). The decreasing number of trees in the Kuala Krai and in the state of Kelantan increases the probability of the occurrence of floods, and this situation also increases the total damage – especially in areas that have more residents and activities. Beyond that, the massive illegal land clearing that led to deforestation has been reported to happen in Gua Musang where the major rivers begin. As a result, Kuala Krai that is in the location where the rivers meet was damaged the worst due to the inability of the rivers to accommodate with the excessive downpour (Lee 2015). In addition to that, a high rate of massive logging has been reported in Kelantan that results in the increase of timber plantation. In 2014 alone, the state had about 162,000 hectors of land classified as timber plantations. The timber plantation areas where the forest has been clear-cut and the logs have been extracted and removed leave the soil condition bare.

2.5.2.2 Waterways and Unmanaged Drainages

Interruption in the waterways, especially in the rivers, has contributed to the flood in Kuala Krai. The massive flood had Kuala Krai most affected due to its geographical location with the district located in middle of the river mouth. Massive land clearing which is resulting from agricultural activities, whether legal or illegal, is one of the factors contributing to the interruption in the waterways. Due to the vanishing number of trees holding the soil, frequent rainfall has made the soils down to the rivers. This, unfortunately, decreases the river's depth and capacity to accommodate the increase of water level in rainy seasons. Due to that, the water levels in the rivers nearby Kuala Krai were recorded above all flood stage. The highest recorded level of the Sungai Galas in Dabong was 46.47 m that stood above the flood stage of 38 m, the highest recorded level of the Sungai Lebir in Tualang, Kuala Krai, was 42.17 m that was above the flood stage of 35 m, and the highest recorded level of Sungai Kelantan was 34.17 m at Tangga Krai, Kuala Krai, also above the flood stage of 25 m (Wan Hussin et al. 2015). Additionally, portal sources have reported that the rivers had been clogged due to the jammed logs from the massive loggings in Gua Musang. Improper management of the logging activities has resulted in a number of logs drifted into the river stream causing the rivers to be jammed. This, however, caused the water levels to rise and block the waterways to flow in the heavy rain situations. The moral hazard due to the poor management of the logging activities increased the risk exposure to floods happening in the future (Kelantan blames illegal land clearing for uncontrolled logging, December 2014). The authorities in Kelantan and especially in Kuala Krai are responsible for managing the drainage in the state. Unfortunately, poor management is one of the factors that lead to the massive flood in Kuala Krai at the end of 2014. Drainages are crucial in channeling excessive water masses due to heavy rainfall which is significant in the Northeast monsoon season. However, the lack of action from the authorities in cleaning rubbish stuck in the drainage areas and widening the drainage systems to coop with the increase of water in rainy seasons has contributed to the massive flood that struck Kuala Krai the worst at the end of 2014. Beyond that, the lack of attention from the authorities regarding the sedimentation from agricultural and domestic activities reduces the drainage capacities in accommodating the increase of water levels in the rainy seasons.

2.5.2.3 Developments

Residential and agricultural development in the district of Kuala Krai is one of the reasons why the flood impacted the district the most. Rapid development done by developers in the Kuala Krai district has increased the damage due to the flood that hit there at the end of 2014. The district of Kuala Krai is located in the middle of where the main rivers in the state of Kelantan meet, i.e., in the river mouth. In conjunction to that, poor drainage maintenance and the increase of development in the waterway area increased the risk for disaster when the flood happened. In addition to that, rapid development without enough monitoring and without any improvements in the drainage and irrigation system increases the impact of flood. Based on the geographical position of Kuala Kari, most of the developments and residents' area are located in flood-prone areas. In addition to that, developments are also done by clearing the land in order to enable buildings and other construction structures to be built in that area. The land clearing for development purposes increases the likelihood of a flood due to less natural resistance to coping with the amount of rainfall in the monsoon seasons. Additionally, some developers do not give proper attention to the drainage system in the development areas. Some drainage systems are designed to channel normal amount of water but not additional quantities of water resulting from heavy rainfall in the northeast monsoon seasons. Long-term neglect in terms of handling drainage systems increases the frequency and severity of flood incidents. Due to this factor, the magnitude of the flood that is common in Kuala Kari increased tremendously, causing massive loss to many people.

2.5.2.4 Covert Application of Weather Modification Technology

Another possible human factor which is often neglected or even completely ignored in cases of environmental disasters is the possibility of covert application of weather modification technology, e.g., in the form of cloud seeding which is suitable to cause flood disasters and has been applied already since the late 1940s (Pook 2001; Rich 2011). One example for this rain-making weather modification technology covered by renowned newspapers like *The Telegraph* and *The Guardian* has been the *Project Cumulus*, where cloud seeding experiments over southern England performed by the British Royal Air Force between 1949 and 1952 were reportedly responsible for the 1952 flood in the Devon village of Lynmouth resulting in at least 34 deaths and the destruction of diverse structures (Pook 2001; Vidal and Weinstein 2001).

Experiments into weather modification didn't stop in the 1950s but have been continued and refined since then, and even covert warfare by cloud seeding, for example, to extend the monsoon season throughout Southeast Asia has already been conducted in the framework of Project Popeye (1967–1972) during the Vietnam war

(Hap 2013). It is obvious that this was in the same period in which also the Malaysia flood from 1972 happened. Further details about the continued development of weather modification tools from the 1970s to the present as well as details about CE Technology will be covered in a later chapter toward the end of this book.

Nevertheless, the launch of a certain climate engineering (CE) project by Western intelligence agencies in 2013 revealed by the "Independent" (Williams 2013) should be shortly mentioned here because it had just preceded the Kelantan flood and subsequent flood disasters in the USA and in Europe (Duclos 2015; Euronews 2016). Officially, this CE project has been designed to investigate national security implications of CE (Williams 2013), but this official reason might also have served as a pretext to refine CE Technology even further for hostile purposes which are presumably only known by those who financed, supported, and conducted this project. Taking these facts into consideration, one of the causes which led to the Kelantan flood in this magnitude might eventually have been covert experiments with CE Technology in Kelantan. The possibility of a covert hostile attack on Kelantan by unknown players with access to CE Technology and a hidden agenda with the objective to harm Kelantan socially, ecologically, and economically cannot be entirely excluded and might be worth to be investigated.

2.5.3 Effect from the Flood Toward the Research Area

2.5.3.1 Losses and Deaths

The direct loss that can be identified in the flood incident in Kuala Krai is the loss and damages of houses and personal belongings. It has been estimated that almost 23,500 people were affected by the flood in Kuala Krai, Kelantan (Banjir Di Kelantan Semakin Buruk, Jumlah Mangsa Meningkat 2014). The monetary loss that has been recorded by the authority in the flood incident has estimated to reach RM200 million (Davies 2015). Additionally, schools, residential areas, infrastructures, buildings, livestock, and public and private properties have been reported to sustain huge damage due to the flood. In addition to that, the number of death total in Kuala Kari due to the increase of water level could not be estimated due to the massive flood. In addition to that, a huge number of residents in Kuala Krai have been reported to have sustained total loss that includes loss of personal belongings due to being washed away by the flood and burglary that happened after the flood had receded. Some of the flood victims have lost their houses and became homeless. The monetary loss that has to be carried personally by the flood victims will cause them problems. Flood victims who did not take up insurance coverage to cover their losses will have difficulties to recover from the massive loss. The cost of recovery will include the cost of repairing their home and other properties. Furthermore, some victims would have to incur cost of reinstating their home due to extreme damages and even full loss of their home resulting from the flood.

2.5.3.2 Diseases, Psychological Effects, and Contaminations

Vector borne disease outbreak has also been reported in the flood incident. The major diseases that have been reported by the news sources are malaria and dengue fevers resulting from stagnant water areas that can change into breeding sites for vectors (WHO 2015). Other diseases that have an outbreak during flood incidents are wound infections, dermatitis, conjunctivitis, and ear, nose, and throat infections (WHO 2015). A survey was done by MERCY Malaysia which yielded that flood victims are exposed to the following four diseases that are common in flood seasons: upper respiratory tract infection (URTI); skin diseases such as fungal infection; loss of diabetic, antihypertensive medications; and lastly musculoskeletal pains (Ibrahim et al. 2015). Moreover, clean water resources have also been damaged due to the flood which makes the situation more difficult for the flood victims to gain clean water resources. Due to this condition, the contaminated water resources increased the probability of epidemic break out in the flood areas. The flood victims also experienced psychological effects such as stress and trauma due to the massive flood. Stress and trauma may occur during and after the flood because of the dramatic changes and the loss of property and – to some extent – loss of the lives of family members. Additionally, stress and trauma may occur due to the communication cutoff with other family members that are located in other places.

2.5.3.3 Economic Effects

Due to the massive flood that struck Kuala Krai, most of the residents in the district are exposed to social effects such as unemployment. The catastrophic flood that hit Kelantan damaged many shop lots, business premises, farms, and lands. This directly impacts the source of employment of most of the residents in Kuala Krai. In addition to that, most of Kuala Krai residents are in the position of facing a loss in income in a long period of time until they will be able to rebuild their damages and losses. Flood victims who earn their living with agricultural activities will incur a higher cost of recovery due to the total loss of their crops or livestock. Over and above that, the authorities will also have to bear massive costs in order to reinstate damaged infrastructures in the district. Additional cost will also be insured by the authorities in terms of supplying food and other life necessaries to the flood victims during the flood and after the flood has receded. Further, scheduled development has to be delayed in order to focus more on the reconstruction and reinstating damages that result from the catastrophic flood. As Kelantan is famous for its small and medium enterprises, it is the main source of income for certain kind of families. But the flood had swallowed their premises. It did not only incur financial loss but also future income. They cannot run the business anymore, and that would be a problem, and it will take time to find new sources of income. In addition to that, huge monetary losses will impact the countries condition. This is because the government will have to spend a big amount of money in order to recover the states that have been hit by the flood. In conjunction to that, the flood that hit will also impact insurance companies. This is because insurance companies are forced to pay the losses that the insured experienced due to the flood.

2.6 Recommendations

2.6.1 Pre-loss Recommendations

2.6.1.1 Disaster Risk Management for Sustainable Development

The recommendations that can be suggested to the target audiences to prevent this from happening again in the future will be through a flood disaster management framework based on the framework that is suggested by the HYOGO Framework and further improved using the Disaster Risk Management for Sustainable Development (DRM-SD) model that is developed by the Centre for Global Sustainability Studies (CGSS), University of Science, Malaysia. The frameworks emphasize four pillars (prevention, preparedness, response, and recovery) (Ibrahim et al. 2015). Governance is an additional pillar in DRM-SD of USM. It is to show how important their role in managing disaster management. Governance can be defined, anything that has to do with the management of the four pillars in DRM-SD of USM, that is, prevention, preparedness, response, and recovery (Ibrahim et al. 2015). The first pillar emphasizing prevention in DRM-SD of USM has stated that construction developers, government agencies, and residents should avoid development and building a house in flood-prone areas, for example, lands that are located in a low area such as under the hills or near to the sea or river. This prevention pillar is suitable to be implemented before flood disaster happens in order to avoid the flood. This is due to the geographic condition, bad weather conditions, and heavy rainfall at the end of the year (Ibrahim et al. 2015). In addition to that, construction companies should design housing that follows the guidelines from Malaysia Civil Defense Department and Fire and Rescue Department of Malaysia. Developers should implement a strong and effective housing structure in order to stay resistant when a flood happens. Developers can also implement a technology from Japan that is called Cross-Wave (News 2015). This device acts as an underground reservoir replacing river stones that is present in the existing catchment. The Cross-Wave can accommodate the rainwater before it is released into the nearer river. Although in Malaysia there is an insufficient amount of green spaces to develop rainwater catchments, the problem can be solved by using a special polypropylene sheets and channeling devices (Ibrahim et al. 2015).

The second pillar is preparedness that is suitable to be implemented before flood disaster happens. The Ministry of Health Malaysia should collaborate with NGOs to set up sufficient medical teams to help transfer and give treatments to the flood victims when the flood strikes. The number of medical teams can be estimated based on the historical database in the nearest hospital. Medical priority should be given to flood victims who are vulnerable, for instance, those who need dialysis, pregnant mothers, babies, and chemotherapy patients (Malaysia 2015). This is important to prevent the patients becoming more ill due to improper treatment and to prevent diseases that are caused by the flood situation. Based on the survey done by the MERCY Malaysia, flood victims are exposed to four common diseases: these are upper respiratory tract infection (URTI); skin diseases, for example, fungal infection; loss of diabetic and antihypertensive medications; and lastly musculo-skeletal pains (Ibrahim et al. 2015).

The third pillar is the response pillar that explains how the responsible department should respond to the situation and make some changes about certain factors that need to be improved. The National Security Council with the cooperation of the Ministry should perform the triage system under the response pillar when handling the victims and be prepared to solve the case.

The triage system is divided into three categories. The red and yellow are categorized as emergencies. They (these categories) should be treated fast and effectively by the emergency departments. The green category represents low high probability so it should be treated in field hospitals or clinics. It is important to take suitable measures to avoid interference in terms of the availability system in hospitals (Ibrahim et al. 2015). The fourth is the recovery pillar which states about the plan to relocate the location of the facilities after the flood and also replace damaged assets. According to current information by relevant sources, the flood in Kelantan at the end of 2014 had badly affected school which made it impossible for many students to start school by the beginning of the year.

In conjunction to the fourth pillar, the Ministry of Education and the State Education department should take measures to repair the damaged schools. The authorities should also move non-waterproof materials such books; wood furniture, for example, tables and chairs; as well as electronics' equipment to a safer place or higher ground before the flood water level exceed the yellow levels. This can also help lower the costs of recovering the damaged items which increase every year. In addition to that, taking these measures can reduce the time of recovery and enable the students to start the learning process as soon as possible (Ibrahim et al. 2015). Other departments should also perform their responsibility according to the case that falls under their scope of jurisdiction.

Other departments should also perform their responsibility according to the case that falls under their scope of jurisdiction. For example, MERCY Malaysia contributed household items including blankets, mosquito nets, eating utensils, stove, and sanitary pads to the flood victims and also distributed some mattresses and pre-packed hygiene items after the flood had receded in the flood area. Beyond that, MERCY Malaysia also contributed school items, school uniform, pairs of shoes, sets of stationeries, sets of writing pads, and school bags to the flood victims (Seringgitsehari, 2015). Authorities, the Department of Social Welfare, nongovernmental organizations, district office, and the National Security Council, can provide, under the recovery pillar, food as the most important requirement for flood victims. Inability by providing clean and sufficient needs to the flood victims will result in starvation and disease and increase the number of death as a post loss effect of the flood (Ibrahim et al. 2015). Beyond that, one might also implement coupon system for food and basic needs in the flood evacuation centers to prevent double receives, thus reducing the possibility to experience a shortage of relief supplies.

2.6.1.2 Adaptation of Japan Fukushima Daiichi Drainage System

Another recommendation that can be implemented by the authorities is adopting the Japan Fukushima Daiichi drainage system to solve this problem. Fukushima Daiichi drainage system is used by collecting underground water and storing it temporarily to check the quality. If the quality is suitable for the process, the underground water will go into thorough treatment process. This is done in order to avoid contaminated water to flow out of the flooded area and create a bigger negative effect. As we know that the flood in Kelantan was also known as the "Red Flood" due to its color changes resulting from the mixture of mud and other elements in the flooded area, we believe that the Fukushima Daiichi drainage system can be implemented in the area to reduce the magnitude of damage done by the flood. Based on the Fukushima Daiichi drainage system model, the water that has undergone the treatment is then soon released into the harbor area located in another location. By adopting this system, the government could consider building a sub-drain system that uses wells which were erected around the reactor and turbine buildings. The sub-drain systems will pump underground water that is flowing in the flooded area out into the fields and reduce the underground water quantity. By implementing the system, flood areas that are filled with water are able to dry faster than using natural methods, and this can also reduce the impact of pollutants from spreading into other areas due to the system also doing water treatment procedures. In addition to that, according to reports, the flood that happened in Kelantan in the last 2 years had difficulty in drying up because the water was unable to seep into the ground due to the presence of other elements preventing it. Taking this situation into consideration, we recommend the implementation of this system in order to reduce the magnitude damage of the flood (UN-Water, 2014).

2.6.1.3 Awareness Programs

The Kuala Krai District Council should take initiatives to promote a flood awareness program by explaining the elements related to the frequent flood and especially in schools because our young generations should know the importance to take care of our environment. This flood awareness program also explains the causes and effects of the flood. This program also explains what the standard operating procedure (SOP) is when a flood happens to prevent something worse happening (Ibrahim et al. 2015). Beyond that, this measure is important to be introduced to the younger education in order to avoid unwanted incidents related to children who do not realize the danger of flood. This is because it is common being reported by the news about children swept away by flood water during a flood. In addition to that, flood awareness should be seriously relayed to adults in order to increase their understanding related to the flood. This is because it is also common to hear in the news that some adults who are traveling got trapped in their cars due to the flood. In order to do so, the Kuala Krai DIstrict Council can organize talks and campaign with the collaboration of the Ministry of Environment Resources and Environment in order to explain the causes, effects, and necessary measures that have to be taken as pre- and post-incidents in order to reduce the risk exposure, severity, and frequency of the flood to happen. The risk of flood can be reduced by organizing a cleanliness campaign and inviting the residents to maintain cleanliness in the drainage systems. Moreover, campaigns such as "care the river, waterways, and drainage system" can also be implemented to increase awareness among the residents in Kuala Krai and encourage them to help reduce the magnitude of flood. Insurance and Takaful companies would also be a good recommendation to be considered wherein these mentioned companies need to notify or aware the residents in Kelantan on the provided or available natural disaster insurance coverage; this notification or awareness can also be conducted through several channels such as creating an awareness campaign or public talks carried to residents in order to expose the insurance benefits and coverage to them. This is because not many of the residents in Malaysia are well aware and many of them have no knowledge on the natural insurance coverage that is available in the market. Further, some residents are still skeptical about the benefits of insurance coverage. In order to reduce their personal losses, they should be exposed to the insurance coverage.

2.6.1.4 Dams Monitorations

Another recommendation that can be implemented as a pre-loss action is through the detailed observations of water levels in dams located in potential flood areas (Pradhan and Youssef 2011). Dams have their own flood-control reservations to which the level of their water should be kept before releasing the water in the rainy season. They have a specific level of water space that the floodwaters can fill. Unfortunately, in some cases, the dam is unable to accommodate with the excessive increase of the waters during the rainy season forcing them to release the excessive water. This is done in order to prevent the dam from experiencing damages that could result in catastrophic damages. However, the release of the excessive waters creates the flood problems. In order to avoid the flood problems from happening, Malaysia should consider implementing dry dam of their own. Dry dam is a dam purposely used to keep the flood water in the rainy season. Rather than releasing it in residential areas and damaging their properties, we could consider channeling the water into the dam. The examples of the dry dams are Mount Morris Dam that is located in south of Rochester, New York, and Seven Oaks Dam that is located in the northeast of Redlands, California.

2.6.1.5 Natural Disaster Insurance Coverage

Another recommendation that we would like to recommend is the purchase of natural disaster insurance. As we know, Kelantan and especially Kuala Krai area is a flood-prone area. In order to decrease the magnitude of loss experienced personally by the victims, they should consider transferring their risk to insurance companies that provide coverage for flood. It is known that Kuala Krai has the high frequency of flood incidents and the severities of the floods are also high. In order to manage the risk, flood victims should consider purchasing insurance coverage in order to compensate their loss. Additionally, the government can also create a government insurance related to natural disaster coverage for those who are unable to pay for insurance sold by the insurance companies. This is because some companies do not provide coverage to natural disaster losses due to the magnitude of loss. In order to help relieve the hardship of the needy, the government can consider creating government insurance coverage.

Insurance can be a disaster risk reduction tool to increase community resilience. Insurance against natural perils such as flooding is useful in providing low-income households and businesses with the right to post-disaster liquidity. Priority 3 of the Sendai Framework has outlined insurance as one of DRR measures to reduce the financial impact of disasters on governments and societies. In a study of the effectiveness homeowners' insurance in mitigating flood losses in the district of Kemaman, Malaysia, Prabhakar et al. (2017) found that the benefits of insurance outweigh its cost. Thus, insurance can lessen the burdens from disasters by securing livelihoods and expediting the recovery process. Currently, the Malaysian insurance market provides coverage against flood through the fire insurance and homeowners insurance. Flood is a standard coverage in homeowners' insurance, while in fire insurance, flood can be covered with an additional premium. Aside from flood, the policy covers losses due to other perils like fire, lightning, hurricane, and windstorm. In addition, a new insurance product that provides coverage for wooden houses was introduced by Allianz in 2013. Aside from property losses, this policy provides emergency cash fund of RM1,000 per loss or in a series of losses during period of coverage against flood and windstorm. Despite the increasing risk exposure due to flood, the market penetration for general insurance in Malaysia is very low which is only 1.7% (Central Bank of Malaysia 2013). This indicates a low awareness on the availability of insurance to cover losses due to flood. There has been an initiative from the Malaysian government to explore the role of insurance in mitigating losses due to disaster including the flood risk. The National Disaster Management Agency (NADMA), an agency established under the Prime Minister's Department to coordinate disaster management efforts, has prepared a proposal to the Cabinet on the establishment of an insurance program to cover losses due to disaster (Utusan Online 2016). Nonetheless, to date, no public insurance program has been announced. Private-public partnership in providing the disaster insurance coverage may be the best solution for the Malaysian case. The low penetration of the general insurance market indicates that the private insurance industry has more capacity to provide coverage due to flood losses. However, properties in the flood-prone area such as Kuala Krai may not be able to obtain coverage due to the restriction in the underwriting procedures. In other words, insurance companies are unlikely to provide coverage for low-probability, high-impact risk exposures. Thus, providing insurance coverage is only possible with government intervention, for example, the UK Flood Reinsurance Program (Flood Re) which received regular financial support from the government through subsidies in insurance premium as well as public recovery aid (Mysiak and Blanco 2016). In this regard, innovation in the financing mechanism of the natural disaster insurance for Malaysia is urgently needed in ensuring effective recovery from disaster losses.

2.6.2 Post Loss Recommendations

2.6.2.1 Government Aid

Flood victims really need financial support from the government to rebuild their destroyed home during flood incident. The government should provide financial assistance to the flood victims. The government should also allocate funds for the reconstruction of settlements. The government should attach greater importance to the matter than to other things. Thus, financial support from the government is needed to rebuild the destroyed settlements. In the 2014 flood, the Kelantan Post- Flood Joint Committee stated the first phase distribution of RM500 aid to flood victims involved evacuated to gazetted relief centers (BERITA 2015). This might not be much, but it helps to relieve the burden on their shoulders. The list of victims and those who received assistance in this incident will be reviewed by the federal and state governments before the list is submitted to the National Security Council (NSC) at the latest on 16 January, and the list will be on display at the District Land Office or the Social Welfare Department (SWD) for review and future actions (Ibrahim et al. 2015). Assistance skills from various fields should be provided to redevelop the site of the incident. In this scenario group of people, experts in the construction and psychology field are needed (Davies 2013). Experts in the field of construction can use their experience and also skills to rebuild the devastated scene so that the area damaged by the flood can be restored quickly. The psychology committee can also provide emotional treatment to the victims. Many flood victims experience emotional distress in the aftermath of a terrible natural disaster. For example, they may not be able to sleep, having nightmares and trauma. Psychologists can help them to overcome their emotional problems in the right way. So, expert assistance in various areas should be provided so that they can help victims who need help (Ibrahim et al. 2015).

2.7 Conclusion

This study attempted to find out whether natural causes and human factors have any relationship to the catastrophic flood in Kelantan or not. However, based on the findings, it can be concluded that both natural and human factors have a significant relationship with the catastrophic flood incident in Kelantan that happened in the last of 2014. The researchers strongly believed that the natural and human factors contributed to the catastrophic flood in Kelantan state. It can specifically be concluded that the natural factors which contributed to the catastrophic flood in Kelantan were due to the northeast monsoon and climate changes. For Northeast monsoon and climate changes, we recommended two ways which are dry dam monitoring and natural disaster insurance coverage. Dry dam monitoring is used to keep flood water in the rainy season. By doing so, the water will go into the dam instead of being released to the residents' area. While the natural disaster insurance coverage used to provide coverage due to natural disaster to avoid that the residents face total lost, the human factors are contributed by land clearing, waterways, unmanaged drainage system, and developments in the research area. Regarding the human factors, it is recommended to establish a sustainable development framework for a disaster risk management. This is one of the flood disaster management frameworks which emphasize the four pillars (prevention, preparedness, response, and recovery) as the most suitable measures that can be taken in this time to manage the flood incidents before and after. Additionally, we also recommended the Fukushima Daiichi drainage system. This system is to provide alternatives to collect the underground water to prevent the water from coming out to the residential area. Last but not least, we recommend an awareness program. This program is to provide knowledge and awareness to the residents about the importance to do take the right kind of actions to minimize risks and to stay safe. The monetary effect of this accident is very costly, and it involved a lot of money to spend in order to rebuild the houses, facilities, and infrastructures. Other effects from this incident were loss of human lives where many families lost their family members, children lost their parents, and parents lost their children. The flood incident also had a negative impact on our economy where many employees lost their jobs and employers lost their companies, materials, and products. In conjunction to the losses, we need a lot of time to restore the current situation back to where it was before the flood happened. As general though, the researchers believe that the former flood victims should also be aware of the importance to opt for insurance or Takaful policy to cover natural disaster accidents in order to manage their personal monetary losses.

In addition to that, responsible authorities with the cooperation of various companies should take awareness steps in order to promote insurance, Takaful products, and coverage to the former flood victims. Last but not least, the researchers also believe that not only former flood victims should be aware about this coverage but also everyone because the risk of flood disasters does still exist. In a nutshell, we could say that the risk of flood disasters cannot be avoided due to the natural condition that is out of human's control, but the frequency and severity of the incident can be lowered in order to control the magnitude of loss. This goes especially for flood disasters which might have been caused by the deliberate human factor. Therefore, independent meteorologists and other experts such as engineers, professional investigators, and diplomats should be consulted by the Malaysian people and the government to investigate whether the magnitude of the 2014 flood might have been the result of deliberate weather modification, e.g., by field experiments in CE Technology.

Actually, it seems rather strange that the Kelantan flood happened just a year after a research program on CE Technology conducted by Western intelligence agencies had been started (see above under 2.5.2.4). In case that it should turn out that the catastrophic Kelantan flood 2014 was part of this CE experiment of Western intelligence agencies at the expense of Malaysia or even a deliberate attack by CE Technology with hostile intentions to damage the people, the natural environment, and the economy of Malaysia in general or Kelantan in particular, one should try to find out who were the responsible actors behind this attack and involved in this attack within or outside Malaysia.

Based on these findings, one might take the necessary legal, political, or diplomatic steps to get damage compensation from those responsible for this attack and to take precautions against this kind of attacks in the future. Further details about the abuse of weather modification and CE Technology (Rich 2011; Schwarz-Herion 2015, with reference to The Royal Society 2009) and possible solutions to this problem will be treated in another chapter toward the end of this book. In any case, future floods of this magnitude like the one in Kelantan can only be prevented if all factors (natural factors as well as negligent and deliberate human factors) are taken into consideration. This should also involve investigations to figure out all causes of the Kelantan flood and to take the necessary measures of problem handling in the framework of crisis management.

References

Ali Khan MM, Shaari NA, Achmad AM, Baten MA (2004) Impact of the flood occurrence in Kota Bharu, Kelantan Using Statistical Analysis. J Appl Sci 14:1944–1951

Ali Khan MM, Shaari NA, Achmad Bahar AM, Baten MA, Nazaruddin DA (2014) Flood impact assessment in Kota Bharu, Malaysia: a statistical analysis. World Appl Sci J 32(4):626–634

Azlee A (2015) Worst floods in Kelantan, confirms NSC. Retrieved on January 5, 2017, from The Malay Mail Online: http://www.themalaymailonline.com/malaysia/article/worst-floods-in-kelantan-confirms-nsc

Baharuddin KA, Abdul Wahab SF, Nik Abd Rahman NH, Nik Mohammad NA, Tuan Kamauzaman TH, Md Noh AY, Abdul Majid MR (2015) The record-setting flood of 2014 in Kelantan: challenges and recommendations from an emergency medicine perspective and why the medical campus stood dry. Malays J Med Sci **22**(2):1–7

BERITA (2015) Distribution of RM500 aid to Kelantan flood victims based on three categories. Flood News. Retrieved on August 11, 2017, from http://english.astroawani.com/flood-news/distribution-rm500-aid-kelantan-flood-victims-based-three-categories-51987

Bernama (2015) Uncontrolled logging in Kelantan the cause of flooding. Free Malaysia Today, Parit Buntar

Butler R (2012) Soil erosion and its effect. Retrieved on June 21, 2016, from MongaBay.com. http://rainforests.mongabay.com/0903.htm

Central Bank of Malaysia (2013) Insurance annual statistics. Retrieved from All News PipeLine on March 30, 2016, from http://www.bnm.gov.my

Crabtree A (2012) Climate change and mental health following flood disasters in developing countries, a review of the epidemiological literature: what do we know, what is being recommended? Australas J Disaster Trauma Stud 1:21–29

D/iya SG, Gasim MB, Toriman ME, Abdullahi MG (2014) Floods in Malaysia historical reviews, causes, effects and mitigations approach. Int J Interdiscip Res Innov 2:59–65
Davies R (2013) Floods list. Retrieved September 30, 2015, from Malaysia Floods Update. http://floodlist.com/asia/malaysia-floods-update, December 4
Davies R (2015) Malaysia floods – Kelantan flooding worst recorded as costs rise to RM1 Billion. Flood list
Department of Irrigation and Drainage Malaysia (2009) Flood management manual vol 1. Department of Irrigation and Drainage Malaysia, Kuala Lumpur
Dictionary, B (2015) Business Dictionary.com. Retrieved October 3, 2016, from flood: http://www.businessdictionary.com/definition/flood.html
Disaster hits Malaysian shores (2015) Retrieved from World Vision Malaysia: https://www.worldvision.com.my/malaysia-flood
Duclos S (2015) Is Texas under attack by the U.S. government for resisting Jade Helm? "Weather Warfare Right Before Our Very Eyes". Retrieved from all news pipeLine on June 14, 2016, http://allnewspipeline.com/Texas_Under_Attack_Resisting_Jade_Helm.php
EuroNews (2016) Rising floodwaters in Germany and France claim more lives. Retrieved from EuroNews online on June 13, 2016, http://www.euronews.com/2016/06/02/rising-floodwaters-in-germany-and-france-claim-more-lives/
FEMA (2015) Flooding and flood risk. Retrieved on October 15, 2016, from https://www.floodsmart.gov/floodsmart/pages/flooding_flood_risks/ffr_overview.jsp
Hap (2013) Operation Popeye: weaponized weather during Vietnam War. Retrieved on March 11, 2017, from Opsecnews on http://www.opsecnews.com/operation-popeye-weaponized-weather-during-vietnam-war/
Human Factors and Ergonomics Society (HFES), (2015) Definitions of human factors and ergonomics. Retrieved from Human factors and ergonomics society: Retrieved on March 11, 2017, from http://www.hfes.org/Web/EducationalResources/HFEdefinitionsmain.html
Ibrahim KA, Koshy KC, Akib NAM, Nor RM, Manaf A, Azhar SNFS, Muslim M (2015) Resolution: Kelantan flood disaster management conference, resolution: Kelantan flood disaster management conference 2015, pp 6–21
Indaramalar S, Soon I (2015) Kelantan flood victims in makeshift shelters at the mercy of the elements. Retrieved on March 27, 2017, from The Star Online. http://www.thestar.com.my/Lifestyle/People/2015/03/27/Flood-victims-in-Kelantan-continue-to-be-at-the-mercy-of-the-elements-in-their-makeshift-shelters/
Khalid MS, Shafiai S (2015) Flood disaster management in Malaysia: an evaluation of the effectiveness flood delivery system. Int J Soc Sci Humanit 5(4):398–402
Lee P (2015) River confluence caused worst floods in Kelantan. Asia One: Singapore Press Holding Portal, Kuala Lumpur
Mahavera S (2015) Malaysia could see more severe floods like in Kelantan, say experts. The Malaysian Insider
Malaysiakini (2014) Flood victims tally tops 200,000, Kelantan worsens. Retrieved on January 2, 2017, from http://www.malaysiakini.com/news/284726
Malaysian Meteorological Department (2014) Weather warning. Retrieved on September 30, 2016, from Heavy Rain Warning. http://www.met.gov.my/in/web/metmalaysia/weatherwarning
Malaysia Mail Online (2014) Number of flood evacuees in Johor rise; no change in Perak, Kelantan and Terengganu. Retrieved on May 12, 2017, from http://www.themalaymailonline.com/malaysia/article/number-of-flood-evacuees-in-johor-rise-no-change-in-perak-kelantan-andtere#rGVVc563TRjpATzq.97
Morris-Oswald M, Simonovic SP (1997) Assessment of the social impact of flooding for use in flood management in the Red River basin. Report prepared for International Joint Commission Red River Basin Task Force, Winnipeg, Canada
Mysiak J, Blanco CDP (2016) Natural hazards and earth system. Science 6(11):2403–2419
Ooi SH, Samah AA, Braesicke P (2011) A case study of the Borneo Vortex genesis and its interactions with the global circulation. J Geophys Res 116(D21). https://doi.org/10.1029/2011JD015991
Permin AHM (1997) A cross-sectional study of helminths in rural scavenging poultry in Tanzania in relation to season and climate. J Helminthol:233–240

Pook S (2001) Deadly flood blamed on RAF rainmakers, Telegraph, 31 August 2001, Retrieved on June 14, 2017, from http://www.telegraph.co.uk/news/uknews/1339046/Deadlyflood-blamed-on-RAF-rainmakers.html
Prabhakar SVRK, Solomon DS, Abu-Bakar A, Cummins J, Pereira JJ, Pulhin JM (2017) Case studies in insurance effectiveness: some insights into costs and benefits. Southeast Asia Disaster Prevention Research Institute (SEADPRI), Bangi
Pradhan B, Youssef AM (2011) A 100-year maximum flood susceptibility mapping using hydrological and hydrodynamic models: a case study. J Flood Risk Manage 4:189–202
Pure Treatment Plant (2017) AWG Glossary of water filtration and water related terms. Retrieved on May 2, 2017, from https://www.purewatertreatment.com/index.php?option=com_glossary&Itemid=0&letter=&page=80
Rich M (2011) Weather warfare. Introduction. Retrieved on June 13, 2016, from New World War online: http://www.newworldwar.org/weatherwar.htm
Sakar SK, Begum RA, Periera JJ, Jaafar A (2013) Addressing disaster risk reduction in Malaysia: Mechanisms and responds. Paper presented at 2nd International Conference on Environment, Agriculture and Food Science (ICEAFS2013), August 25–26
Seringgitsehari (2015) Water, sanitation and hygiene (WASH) Programs. Retrieved on May 2, 2017, from http://seringgitsehari.org.my/watersanitation-and-hygiene-wash-programs/
Schwarz-Herion O (2015) Urgent ecological problems for the NSP. In: Schwarz-Herion O, Omran A (eds) Strategies towards the new sustainability paradigm. Managing the great transition to sustainable global democracy. Springer, Cham, pp 121–140
Tan F, Lim HS, Abdullah K, Yoon TL, Holben B (2015) AERONET data–based determination of aerosol types. Atmosph Pollut Resea 6(4):682–695
The Royal Society (2009) Geoengineering the climate. Science, governance, and uncertainty. Science Policy Centre, London
Thornes JE, Pope FD (2014) Why do we need solutions to Global Warming? In: Hester RE, Harrion RM (eds) Issues in Environmental Science and Technology, vol 38. Geoengineering of the climate system. The Royal Society of Chemistry, Cambridge
UN-Water (2014) A post-2015 global goal for water: synthesis of key findings and recommendations from UN-Water. Retrieved on May 2, 2017, from http://www.un.org/waterforlifedecade/pdf/27_01_2014_unwater_paper_on_a_post2015_global_goal_for_water.pdf
Utusan Online (2016) Penubuhan NADMA hadapi banjir beri kesan baik. [online] Available at: http://www.utusan.com.my/berita/nasional/penubuhan-nadma-hadapi-banjir-beri-kesan-baik-1.175051. Accessed 27 Mar 2016
Vidal J, Weinstein H (2001) RAF rainmakers caused 1952 flood. Retrieved on June 14, 2016, from the Guardian online: http://www.guardian.co.uk/uk/2001/aug/30/sillyseason.physicalsciences
Vincent RK (1997) Fundamentals of geological and environmental remote sensing. Prentice Hall. Inc., Upper Saddle River
Wan Ahmad WI, Abdurahman SM (2015) Kelantan flood 2014: reflections from relief aid mission to Kampung Kemubu, Kelantan. Mediterr J Soc Sci 3(2):340–344
Wan Hussin WNT, Nor Hidayati Zakaria NH, Ahmad MN (2015) Knowledge sharing and lesson learned from flood disaster: a case in Kelantan. J Inf Syst Res Innov 9(2):1–10. http://www.who.int/hac/techguidance/ems/flood_cds/en/
Williams R (2013) CIA backs $630,000 study into how to control the global weather through geoengineering. Study part-funded by the CIA to investigate national security implications of geoengineering. Retrieved from The Independent online on May 14, 2017, from http://www.independent.co.uk/news/world/americas/cia-backs-630000-study-into-how-to control-global-weather-through-geoengineering-8724501.html
World Health Organization (WHO) (2015) Flooding and communicable diseases fact sheet, Humanitarian Health Action. Retrieved on December 15, 2017, from http://www.who.int/hac/techguidance/ems/flood_cds/en/
World Weather Online (2012) Wet or dry season? Retrieved on June 15, 2017, from http://www.worldweatheronline.com/Kuala-Lumpur-weather-averages/Kuala-Lumpur/MY.aspx

Chapter 3
The Effects of Climate Change Risks on the Mud Architecture in Wadi Hadhramaut, Yemen

Mazen Ibrahim Al-Masawa, Norlida Abdul Manab, and Abdelnaser Omran

Abstract This study aims at studying the effects of climate change risks on the mud architecture in Wadi Hadramout, Yemen. The intention of this study was to contribute to mud architecture in Wadi Hadhramaut by identifying and addressing the effect of expected climate changes on mud architecture building to maintain this unique architecture and human heritage toward sustainable development. This study has concluded that climate change scenarios in Yemen, especially in Wadi Hadhramaut, indicate that a significant increase in the rated temperature in that region will appear in the fifth decade of this century. The scenarios also confirm on the expected increase in precipitation, which increases the probability of exposure to the mud architecture in the region to multiple risks and disasters that could reach the threat of the disappearance of villages and cities in Wadi Hadhramaut which contain more than 90% of the components of buildings currently made from mud architecture.

Keywords Climate change · Mud architecture · Risks · Wadi Hadhramaut · Yemen

3.1 Introduction

The impact of natural hazards due to climate change risks on buildings has gained the attention of the building industry as a result of the increasing environmental loss due to hazard events devastating the built environment around the world (Wei et al. 2016). There are multiple international and regional efforts in this regard, but the most prominent would be the climate change treaty (UNFCCC). It was drafted in 1992, and its implementation began within 2 years. Kyoto Protocol in 1997 caused

---

M. I. Al-Masawa · N. A. Manab · A. Omran (✉)
School of Economics, Finance & Banking, Universiti Utara Malaysia, Sintok, Kedah, Malaysia
e-mail: naser_elamroni@yahoo.co.uk

A. Omran, O. Schwarz-Herion (eds.), *The Impact of Climate Change on Our Life*, https://doi.org/10.1007/978-981-10-7748-7_3

an international worrisome of this phenomenon and its causes. However, it was only implemented in 2005 and still suffers from numerous obstacles even though the world is facing temperatures it has not faced in over 10,000 years (Parry et al. 2007).

In Paris, on February 9, 2007, the international meeting was held that included the latest reports of the governmental group assigned to study and monitor climate changes (IPCC 2014). During that meeting about 2500 scientists had confirmed the human responsibility in the changes of climate, and these changes will result in changes in rainfall, snow melting, and the rising of the sea level, which can lead to risks and catastrophes. Therefore, fighting climate change and its effects has surpassed caution to prevention. It is clearly becoming that the climate is changing as a result of anthropogenic activities. The Intergovernmental Panel on Climate Change (IPCC) Fourth Assessment Report unequivocally links climate change to anthropogenic activities, primarily the burning of fossil fuels and changes in land use (IPCC 2014). Today more than half of the global population live in cities. The types of threats that cities are to face can broadly be characterized as follows; in the short to medium term, events such as storms, storm surges, floods, landslides, and wildfires are likely to increase in frequency and severity in the future.

In the medium to longer term, rising sea levels will contribute to coastal inundation.

In the longer term, changes in weather patterns will impact agriculture and biodiversity.

Cities will need to consider how such changes will impact urban development. Cities that will experience warmer weather may have to consider how to increase shade in urban areas. Cities that are likely to experience more intense precipitation events will need to consider their drainage systems and flood defenses. Coastal cities will need to consider the long-term impact of sea level rise and whether additional coastal defenses will be required. A changing climate will also effect tourism patterns as visitor numbers are likely to change throughout the year as people respond to changed climate conditions (O'Brien et al. 2015).

Wadi Hadhramaut is characterized by mud architecture building which is a part of its cultural identity (Warner 1931). However, the possible climate changes are posing its biggest threat, and so, participation in its architectural preservation and rehabilitating it to cope with the climate and environmental changes are of utmost importance (Petzet and Koenigs 2015). It is important to the nations that decided to actively participate in humanitarian aid, on the cultural and historical levels, in a way that reflects its privacy, identity, and heritage in the human space. The mentioned process of participation needs accumulative expertise in the field and all that it represents of professional constants that are currently practiced in construction, as well as knowledge of common beliefs, social, economic, and environmental conditions, as they all directly affect the architecture (Bukair and Al-Madhaji 2015). Therefore, it is important for those who are interested to get knowledge about the identity of this mud-building heritage in Wadi Hadhramaut and its development. In addition, they must possess the knowledge to use the traditional building materials as well as the modern ones, such as moisture and water isolating materials, in a way that can maintain the historical and cultural value of mud buildings and achieve harmony with the neighboring modern and current architecture, as well as the future climate changes (Jerome 2015). Many conducted studies by O'Brien et al. (2015), Parry et al. (2007), Petzet and Koenigs (2015), Schipper et al. (2016), Stewart and Deng (2015), Wei et al. (2016), and Woodward et al. (2014) gave great attention to the study of these climate changes because of their effects on the economic and natural resources. However, these effects and their threats on the urban constructions, especially, in arid areas, are not given the required attention; so that, it is important to assess multifaceted risks of climate changes associated with buildings, especially mud architecture buildings in Hadhramaut. The intention of this study was to contribute to mud architecture in Wadi Hadramout by identifying and addressing the effect of expected climate changes on mud architecture building to maintain this unique architecture and human heritage toward sustainable development.

3.2 The Ongoing Scenario

The phenomenon of climate change is a fact supported by scientific evidence and daily occurrences on the regional and international levels (Chandrasekar and Krishnamurthy 2010). Indeed, the IPCC finds that economic losses from climate- and weather-related extreme events have increased since the second half of the twentieth century. Those increases are to a great extent due to increased exposure of people and economic assets (IPCC 2012). There is also a high confidence that settlement patterns, urbanization, and change in socioeconomic conditions have played a significant role. Thus, it is no longer possible to speak about disasters, development, and climate change, either adaptation or mitigation, in isolation of each other (Schipper et al. 2016). However, the efforts in this matter are still not enough. The third evaluation report of the committee on climate change (IPCC) determined that the problem of climate change might be worse than what was feared before. This was confirmed again in 2007 in the second report of the governmental committee in charge of climate change studies. The latter report stated that an increase in heavy rainfall and floods, especially in dry areas of the third world countries, is likely to happen. Previous studies by Brandstedt (2013), Carter et al. (2015), Costa et al. (2016), Hay and Mimura (2010), IPCC (2012, 2014), Jerome (2009b), Koellisch (2015), Noman (2005), O'Brien et al. (2015), Parry et al. (2007), Petzet and Koenigs (2015), Schipper et al. (2016), Stewart and Deng (2015), Wei et al. (2016), and Woodward et al. 2014 have addressed several aspects of effects of climate change on natural resources, such as water, soil, and vegetation, and on economic aspects, such as agriculture.

The Arab region is facing and will face climate change risks and effects just like the other areas. These risks might affect the climate systems in the forms of heavy rainfall, tornados, floods, an increase in temperatures, and high levels of dryness

Fig. 3.1 A town to the east of Tarim city in Wadi Hadhramaut was destroyed, losing approximately 100 buildings during 2008 flash floods in Hadhramaut (*Source*: The authors)

that can lead to catastrophes and can wipe whole cities, displace big numbers of people, and destroy architectural heritage, which is still full of life and is appropriate for its environment (Al-Saqqaf 2002). For instance, about 5000 mud buildings in Wadi Hadhramaut were damaged and 200 people died in October 2008 due to the catastrophe which caused by heavy rainfall and floods (Jerome 2009a), which emerged many inquiries and questions about mud building that some people are losing faith in it and are turning towards concrete building, so we need to maintain and development of urban construction mudflat coupled with sustainability in order to enable coming to live in this ancient qualities and his distinctive valley generations has adapted according to the requirements of the age and needs. It is important in identifying and adopting with potential climate changes to maintain and sustain this mud architectural civilization and global human heritage immortal.

Although mud brick is considered one of the oldest construction materials, engineers and builders do not have enough information about its mechanical properties. Also, there is no accurate design code to follow before construction (Acosta et al. 2002). There is little support for scientific research, both basic and applied, that relates specifically to the technological and cultural aspects of mud building (Avrami et al. 2008). As well Schroeder (2016) confirmed that research and standardization in this kind of mud architecture are still a vast and complex field. Therefore, this kind of mud architecture projects needs further research. Specifically, the cities and regions of Wadi Hadhramaut in Yemen have a great mud architectural, urban, and cultural heritage which is a part of its cultural identity (Bukair and Al-Madhaji 2015). These cities have been able to provide different humanitarian requirements for their inhabitants over time (Al-Madhaji 2010). However, there are some risks due to climate changes affecting the sustainability of this unique pattern of construction that has been able to last for centuries challenging all the factors, difficulties, and challenges (Fig. 3.1). Accordingly, there is a need to identify the effects of climate changes on the mud architectural building in Wadi Hadhramaut, to help in preservation of this unique architecture building.

3.3 Risks of Climate Changes on Mud Architecture Building in Wadi Hadhramaut

3.3.1 Background of Mud Architecture

Mud architecture is the oldest building material known, with documented cases of the use of mud bricks since Mesopotamia around 10000 BC. Mud construction exists throughout most of the world in diverse cultures, and for some countries, it continues to be the primary process of construction. Dethier (1986) and Coffman (1990) discussed that nearly 30% of total population of the world still resides in mud houses (Fig. 3.2). However, nowadays it is thought that this percentage corresponds to a much larger distribution (Costa et al. 2016).

Currently, almost 50% of the world's population lives in earth-based dwellings (Avrami et al. 2008). It is one of the most popularly used building construction materials in Europe (Sharma et al. 2016). For instance, this kind of construction can also be found in Germany, France, or even the UK that has an excess of 500,000 earth-based dwellings. Earth construction has also increased substantially in the USA, Brazil, and Australia mainly due to the sustainable construction agenda, in which the earth construction assumes a key role (Pacheco-Torgal and Jalali 2012).

3.3.2 Natural and Climatic Characteristics of Wadi Hadhramaut in Yemen

Yemen, with a total area of 527.970 km^2, is located in the Middle East at the south-south-western edge of the Arabian Peninsula between 120 and 190 north of the Equator and 420 and 550 easts of Greenwich. The country is bordered by Saudi Arabia

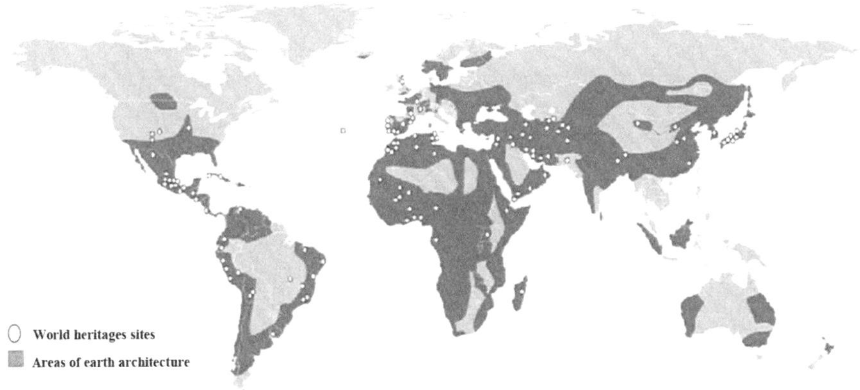

Fig. 3.2 Earth construction worldwide (CRAterre) (*Source*: Costa et al. 2016)

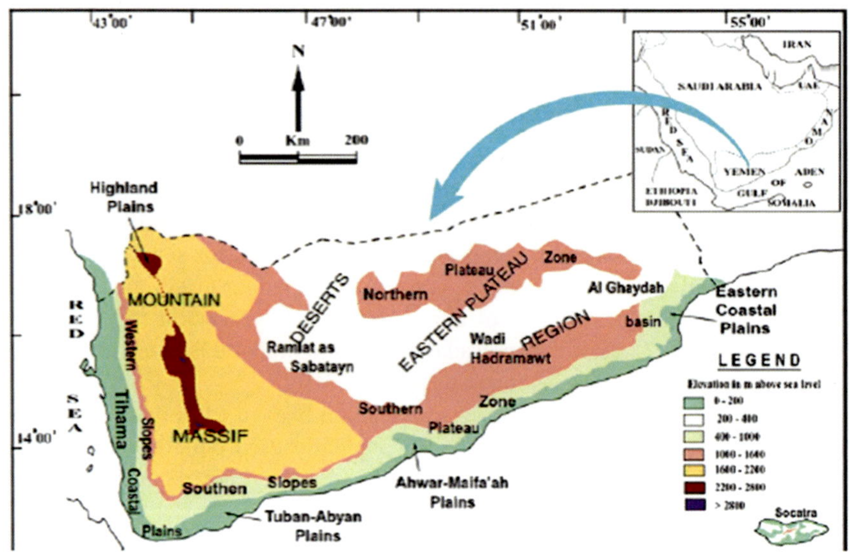

Fig. 3.3 Map of Yemen (*Source*: Al-ameri et al. 2014)

to the north, Oman to the east, the Arabian Sea and the Gulf of Aden in the south, and the Red Sea in the west. The topography of Yemen varies widely. The country's mountainous interior is surrounded by narrow coastal plains to the west, south, and east and by upland desert to the north along the border with Saudi Arabia (Buchalter 2008). The country can be divided according to the altitude and geomorphology into five main regions (Al-Ameri et al. 2014): the coastal plains, the Yemeni mountain massif, the eastern plateau region, the Yemeni islands, and the desert (Fig. 3.3).

Topographical variations of Yemen give rise to a wide range of climatic conditions. In general, the climate of Yemen could be classified as semiarid to arid climate. Temperatures are very high in Yemen, particularly in the coastal regions. The eastern and southern coastal plains are characterized by a high temperature that reaches 42 °C and drops down to 25 °C. Temperature drops down gradually toward higher elevations with an average gradient of 0.6 °C per 100 m difference in elevation to reach 33 °C as a maximum and 20 °C as a minimum. The temperature in the Yemeni highlands drops in the winter and reaches 0 °C. The humidity is very high on the coastal plains (up to more than 80%), whereas it goes down toward the internal parts where it reaches its minimum rate in the desert areas at around 15% (Van der Gun et al. 1995). The rainfall of Yemen depends on two main zones, the Red Sea Convergence Zone (RSCZ) and the monsoonal Intertropical Convergence Zone (ITCZ). The RSCZ, whose influence is most noticeable in the west of the country, is active from March to May and to some extent in the autumn, while the ITCZ reaches Yemen in July to September, moving north and then south again so that its influence lasts longer in the south. Both the RSCZ and the ITCZ produce precipitation in convective storms of high intensity and limited duration and extent, but the

Fig. 3.4 Part of Shibam city in Hadhramaut showing different kinds of mud architecture buildings (*Source*: The authors)

ITCZ storms have a larger areal extent than those of the RSCZ (Farquharson et al. 1996).

The Hadhramaut in the eastern part of Yemen is arid and hot, and the humidity ranges from 35% in June to 64% in January (Al-ameri et al. 2014). Historians, mostly, agree that the significance of Hadhramaut since ancient ages goes back to its geographical situation because it is one of the largest fertile valleys in the Arabian Peninsula (Bukair and Al-Madhaji 2015).

The study area is located in the southern part of Yemen; Wadi Hadhramaut lies between latitudes 14°45′ and 16°30′. The longitudes range between 47°00′ and 49°30′ (El Tahan and Elhanafy 2016). It is connected with south coastal shore with an outlet through Wadi El-Ain, Wadi Do'an, and Wadi Adem; the latter connects Hadhramaut with the nearby Shabwa region. Among the main cities in the mentioned Wadies are Horaidha (at the entrance of Wadi Amd), Al-Hajrain, Al-Khoraiba, Seef (in Wadi Do'an), and Kanina in Wadi Hajer. As for main cities in Inner Wadi of Hadhramaut, they are Shibam, Seiyun, and Tarim (Boustead 1960) (see Fig. 3.4). Hadhramaut is situated 165 km far from the Arabian Sea. It extends between east and west on 16° starting from the eastern edge of Ramlat Al-Sabb'atein; the widest part of it is 15 km, moving toward east to reach its narrowest part 2 km at the area nearby Tarim; from there it closes at Messenger Hood PBUH area which moves southeast to the sea. It ends at this point where it changes its direction and has another name called Al-Masila. Hadhramaut extends on 900 km of lands connected with other smaller valleys that come from northern and southern mountains

(Shamshir and Al-Mashhoor 2009). The nature of Hadhramaut being adjacent to water sources, and, in addition, its fertile soil that makes it easy to grow crops and produce mud bricks are all features that make the area as one of the most convenient areas for mud architecture building (Jerome et al. 1999).

The climate in Wadi Hadhramaut is hot and dry in summer and tends to be cold in winter. The highest average summer temperature is about 35 °C, while the average winter temperature is only 19.7 °C. July is the warmest month, while January is the coldest. Rainfall is rare; the average annual rainfall is 100 mm, except for the higher parts. Although rainfall is rare, severe floods may follow rainfall events causing many damages (El Tahan and Elhanafy 2016). The relative humidity is very low with an annual average ranged between (32–63%); and in the winds are northeasterly during the winter season, while the southwest wind prevails in the summer. It can also be stated that the local winds which called in the local language as "Al-Samoum" can cause sand and dust storms (a meteorological phenomenon common in arid and semi-arid regions) that later could cause a problem in decreasing in seeing the ways and directions between 100–200 meters. (Al-Saqqaf 2002).

3.3.3 Mud Architecture in Hadhramaut, Yemen

Yemeni civilization has been prescribed as the civilization of mud, and that's because Yemeni man has been able since ancient times to make use of and adapt to his land in magnificent mud architecture which represents his ability of creation and innovation, the ability that has been immigrated globally with him to Arab and Muslim countries. The greatness of the distinct Yemeni heritage clearly shows in preserving its unique distinction and the way of construction that survives various natural factors like erosion and climate factors (Ingrams 1935).

The techniques and methods followed by grandfathers by using integrated solution for the environment are still alive and evident everywhere; starting with the construction of roads by accurate engineering methods in transforming and harvesting waters and identifying floods paths and ending with creating the traditional ways of construction that have been used to be followed in cities and villages. They were accurate in making use of the space without affecting the farmlands, using local materials in the construction of houses that are various in size and shape and harmonious with nature and climate. Climatic variations and availability of local materials have had distinct effects on vernacular architecture, which is apparent in its climatic design approaches and the use of suitable, locally available materials and construction methods (Al-sabahi 2005).

Yemen is a perfect case that contains a variety of the methods of constructing with mud architecture building, where these methods differ from an area to another according to different aspects. One main aspect would be the geography and the type of land. This aspect justifies the difference between the methods of mud constructions in different regions such as Sanaa and Hadhramaut (Muhlberger 1998).

Fig. 3.5 Shibam: city of the oldest skyscrapers which is a World Heritage Site (*Source*: The authors)

Mud building construction in Wadi Hadhramaut has been an active regional industry for hundreds of years. It has been historically bound to social structures which have allowed the development of these traditional building techniques and apprenticeship processes as well as created institutions that have supported the craft in the Wadi Hadhramaut for millennia. The enterprises here serve as an example of how regions have been able to capitalize on local raw materials, skills, and labor practices to define the built environment. Today, the mud brick firms in the Wadi Hadhramaut still operate by applying their highly specialized skills. In this way, the process and skills involved have been resilient over time (Mehta 2009). In Wadi Hadhramaut, the mud civilization is classified as unique and distinctive. The mud buildings which had been built for approximately 300 years (Figs. 3.5, 3.6, & 3.7), are still inhabited and well-constructed (Bukair and Al-Madhaji 2015). Damluji (1986) mentioned two major features of mud architecture in Wadi Hadhramaut that make it unique in the world:

First: cities and districts that have been built for 300 years and more are still existing and inhabited till today. During those years, construction depending on same old techniques has not been stopped; builders use the same ways, and no change has been witnessed nor the techniques of construction or construction materials. The sustainable use of local traditional materials has shown, lately, the possibility of continuing local construction depending on rich natural resources; this enhances the richness of Hadhrami architecture and its convenience with the new age, in comparison with the architecture of other countries that have already uprooted their rich traditional architecture and replaced them with inconvenient

Fig. 3.6 Al-Mihdar Mosque in Tarim city in Wadi Hadhramaut, its minaret is the longest mud building in the world (*Source*: The authors)

Fig. 3.7 The Sultan's Palace in Seiyun, the largest mud building in the world (*Source*: The authors)

modern materials to climate, social objectives, or economic circumstances that are present.

Second: this architecture is unique in method and speciality because of what has been accomplished by mud-bricks buildings were raised up to 30 meters high and many elements have been involved among them arches and domes (Damluji 1986). Currently, the important issue that although mud architecture in Wadi Hadhramaut has been able to survive for hundreds of years but recently affected by many different risks which threaten it at the local, regional, national, and international levels (Baeissa 2014).

3.3.4 Climate Change Scenarios in Hadhramaut

Based on the final report prepared by the Panel for the Preparation of Climate Change Scenarios in Yemen at the Public Authority for the Protection of the Environment, which dealt with climate change and its impacts in Yemen until 2050, the two main scenarios related to the study area, which fall within the internal eastern plateaus, namely, drought scenario and humidity scenario, heat and rain, can be summarized as follows (Wiebelt et al. 2011).

3.3.4.1 Drought Scenario

The report of the International Food Policy Research Institute (IFPRI) entitled Climate Change and Floods in Yemen which results from the spatially downscaled climate projections shows that temperatures are expected to rise over their baseline counterpart under both the CSIRO and the MIROC Global Climate Model (GCM) scenarios. However, the variation in temperatures over their baseline equivalents both minimum and maximum differs under the CSIRO and the MIROC scenarios (Fig. 3.8).

With regard to the precipitation, the maximum rainfall during the winter season ranges between 14–18%, and it ranges between 16–28% during spring season while in the summer, it ranges between (5–6%). As per the maximum increase of the annual rainfall rate, it ranges between (9–11%) than in its current condition.

3.3.4.2 Moisture Scenario

A report issued by the International Food Policy Research Institute (IFPRI) entitled Climate Change and Floods in Yemen shows the average monthly rainfall (in millimeters) of the CSIRO scenario which roughly follows the baseline (refer to Fig. 3.9). However, the efforts which carried out by MIROC scenario predicts an increase in rainfall from June to October across Yemen. From October to December, rainfall under the MIROC scenario is below that predicted under the baseline.

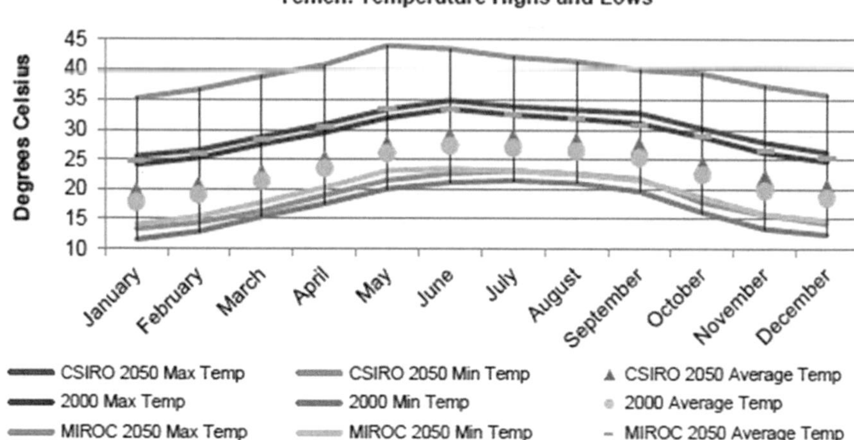

Fig. 3.8 Average monthly temperature in Yemen (degrees Celsius) (*Source*: Wiebelt et al. 2011)

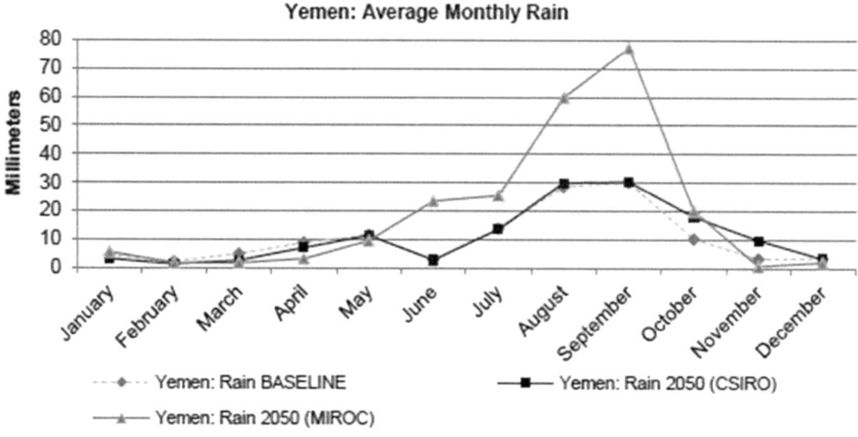

Fig. 3.9 Average monthly rainfall in Yemen (millimeters) Millimeters (*Source*: Wiebelt et al. 2011)

This pattern of variation (or lack thereof for the CSIRO scenario) is consistent across all of Yemeni regions except the Upper Highlands, where the rainfall predictions under the CSIRO scenario are significantly lower than their baseline equivalents (Wiebelt et al. 2011).

This scenario indicates that increasing the extremes of rainfall in the region can reach a ratio of 1% only during the winter season, while it will increase up in the summer with a range rate between (94 and 188%), and it might be increasing more at a ranged rate between (0140–230%) during the fall. On this basis, the annual average rainfall rates will gradually be increasing with percentages more than (84% to 114%) within the internal zone and the eastern plateau which are located in the area of this study (Ali 2012).

Fig. 3.10 The geographical nature of the buildings in the villages and cities of Wadi Hadhramaut (*Source*: The authors)

3.3.5 Potential Risks Due to Climate Change on Mud Architecture

Regardless of the degree of acceptance or reservation is shown by previous scenarios regarding the expected climate of the region to climate change by 2050, the indicators that change began to show clearly in the region. In December 2005, Al-Saqqaf (2002) observed the amount of rainfall in different parts of Hadramout province and in the city of Mukalla in particular (Fig. 3.10). It is a rare phenomenon because rainfall happended during unexpected season and the rainfall had contiuned for about 45 min in Wadi Hadhramaut in the same year which is dangerous phenomenon on mud architecture in the region. Also, there were strong winds in Wadi Hadhramaut in early June 2006, which has not been seen for decades. Some people interested in the weather reported that Socotra had a strong wind and heavy rain at the beginning of June, an unusual phenomenon in this period on the island, which usually comes at the end of June to August each year less than what happened recently. However, if the forecasting of the previous studies on the regional climates which are predicted that for what will be happening in the next coming 45 years especially things related to the rain's issue, then severe disasters and risky matters will happen in the villages and cities located within Wadi Hadhramout. This might later cause the disappearance of these entire cities. A good learned lesson of what happened in Oman lately due to Hurricane Gonu must be taken into serious considering as it gave a clear indication for us in appearing more risks issues due to these climatic changes.

Fig. 3.11 Villages in Wadi Hadhramaut were some of the worst affected by the flash flood in October 2008 (*Source*: The authors)

Moreover, what happened on October 23–24, 2008, approximately 50 cm of rain fell in a 40-hour period in a location that normally receives 7.5 cm annually. The resulting flash flood devastated the Wadi Hadhramaut in Yemen, and the Hadhramaut and Mahra governorates in general (Jerome 2009a). The flash flood was caused by a monsoon that came late in the season and was scheduled to let loose along the coast of the Gulf of Aden. Preparations were made for the storm along the coast; however, the storm unexpectedly veered inland, unleashing itself on the unprepared Wadi. Not only was the intensity and length of the storm unusual, but also the fact that it rained all over the Wadi Hadhramaut and its tributary valleys simultaneously (Figs. 3.11, 3.12, and 3.13). It was the first time that flood waters from the Wadi Do'a reached the main Wadi in 40 years (Jerome 2009b).

3.3.5.1 Increased Rainfall

Mud architecture usually does not have a strong resistance for heavy rain and continuing dropping the rain frequently will be leading to demolition of some mud buildings especially those with less proetction and without surface roof. In addition, there is an urgent need to pay serious attention to the drainage channels and openings of the rainfall from the houses' roofs especially that those designed from mud

Fig. 3.12 An aerial view shows the flood waters following two days of storms in Wadi Hadhramaut on October 2008 (*Source*: The authors)

Fig. 3.13 Ninety-year-old Yemeni man walks in front of his family's collapsed houses in Wadi Hadhramaut on October 27, 2008 (*Source*: The authors)

Fig. 3.14 Using lime in plastering the exterior of the houses at the most areas in Wadi Hadhramaut by rich families (*Source*: The authors)

in Hadhramaut Valley give the roofs (Royom) importance in design and are distributed in more than the role of importance in the summer nights. Homes will receive larger amounts of rainwater, which if left for a while without a quick discharge will expose the house to be quickly demolished (Fig. 3.14). Mud architecture in Hadhramaut Valley has developed methods to resist the impact of rainfall and avoid the negative clay material, by covering the top of the mud house or all with ash and light (Tarqah), this phenomenon is found only in some homes of the wealthy who have the financial means, while most of the houses of the villages and cities of the valley, whether old or newly built, are all clay and exposed to such hazards in the event of increase and frequency of rainfall in the region (Jerome et al. 1999).

3.3.5.2 Heavy Flooding and Disasters

Increased rainfall means flooding, and torrents are likely to occur, especially with the apparent violation of construction on the streams of torrents recently. Even the methods of discharging flood waters and distribution of existing water channels by adopting less rainfall can occur in the future, and this will lead to the destruction of entire areas and cities. In fact, what was happened as a disaster in Wadi Hadhramaut in 2008 is one the greatest examples and indicators of the seriousness of heavy flooding and disasters on the cities and villages of Wadi Hadhramaut, which lead to

Fig. 3.15 Collapsed of mud buildings are seen surrounded by flash flood in a town in the Yemeni province of Hadhramaut on October 2008 (*Source*: The authors)

about of 5,000 mud building damaged and around 200 people died (Jerome 2009a) (see Figs. 3.15, 3.16, and 3.17). On the other hand, the emergence of the effects of climate changes in addition to other factors has affected the increasing trend toward the architectural architecture alien to the environment and not compatible with the natural and social conditions of the region, exposing the region to the loss of architectural heritage that characterizes privacy and achieved architectural security and environmental harmonization (Nahal 2010).

3.4 Conclusion

It is apparent from the foregoing that climate change and its manifestations have become a scientific fact, and their indicators and impacts on the global and regional levels have begun to appear here and there. According to the most recent data issued by the Intergovernmental Panel on Climate Change (IPCC), it is emphasized that climate change and its effects can be more broad and clear and worse than expected. Therefore, the potential risks of climate change on mud architecture can occur.

This study has concluded that climate change scenarios in Yemen, especially in Wadi Hadhramaut, indicate that a significant increase in the rated temperature in that region will appear in the fifth decade of this century. The scenarios also confirm on the expected increase in precipitation, which increases the probability of exposure to the mud architecture in the region to multiple risks and disasters that could reach the threat of the disappearance of villages and cities in Wadi Hadhramaut

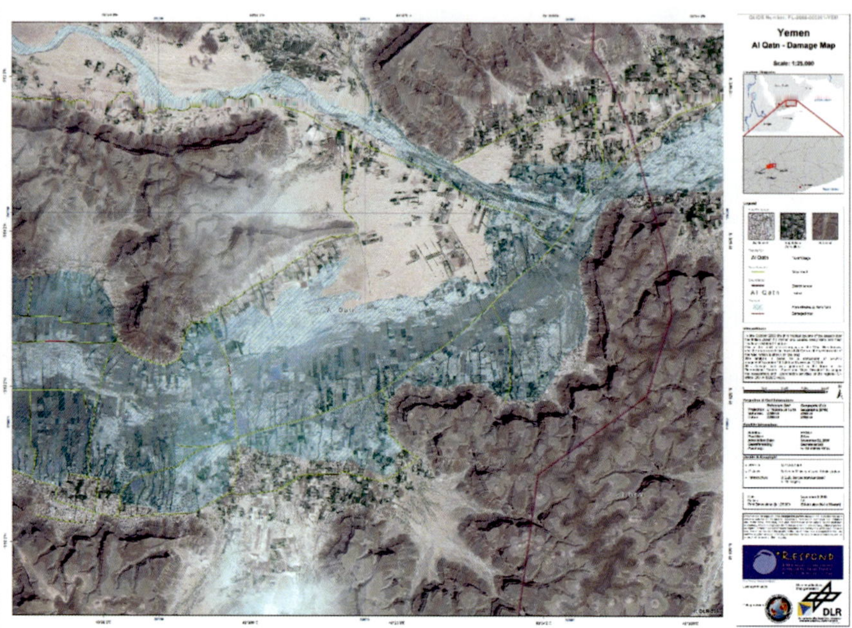

Fig. 3.16 Satellite view of Al-Qatn city during the flash floods on October 23–24, 2008 (*Source*: Jerome 2009a)

Fig. 3.17 Satellite view of Seiyun city in Hadhramaut during 2008 flash floods which is built mostly outside of the flood plain (*Source*: Jerome 2009a)

which contain more than 90% of the components of buildings currently made from mud architecture.

The study shows how this possibility can be achieved through an uncoordinated relationship with the climate changes in the region, especially the heat and rain elements of these changes, and the components and characteristics of mud architecture buildings. The study will help policy makers in government firms in Wadi Hadhramaut, Yemen, to evaluate the projects of mud architecture building and look for climate change adaptation and risk prevention and mitigation to maintain mud architecture buildings in Wadi Hadhramaut toward sustainable development of this unique civilization and immortal global human heritage. Therefore, the study emphasizes the following recommendations:

- Establishing national and regional strategies in order to reform the practical procedures and mechanism to deal with the problems and minimizing the risks on urbanizations.
- Introducing the effects of climate changes on mud architecture buildings in Yemen, especially in Wadi Hadhramaut within the components of the National Program for Adaptation to Climate Change (NAPA), which is supervised by the General Authority for Environmental Protection in Yemen.
- Emphasizing the significance of civil planning and studying the drainage of rainwater properly and banning distributing lands in the paths of floods.
- Exploiting technical expertise and grants from industrialized countries to developing countries on the issue of adaptation to changes through the preparation of programs and the implementation of engineering projects to improve the resistance of mud architecture on the effects of climate change and to maintain the existing mud architecture (e.g., finishing the heads of mud buildings in Wadi Hadhramaut by cement and lime as one of the quick practical solutions).
- Assigning a team of specialized engineers to develop appropriate engineering solutions to enable clay architecture to resist the effects of climate changes on different elements of mud architecture in various villages and cities of Wadi Hadhramaut in Yemen.
- Establishing and activating the role of the Center for Mud Architecture and Disaster Management to perform its duties in protecting mud architecture and developing it.

References

Acosta A, Iglesias I, Aineto M, Romero M, Rincón JM (2002) Utilization of IGCC slag and clay steriles in soft mud bricks (by pressing) for use in building bricks manufacturing. Waste Manag 22(8):887–891

Al-ameri A, Schneider M, Abu Lohom N, Sprenger C (2014) The hydrogen and oxygen isotopic composition of Yemen's rainwater. Arab J Sci Eng 39(1):423–436

Ali W (2012) Governance of climate change in Yemen. Sana'a, Yemen. Retrieved from http://climateresponse.co.za/sites/default/files/Governance.pdf

Al-Madhaji MS (2010) Technology and sustainability in the old cities Meet the requirements – Urban improve – preservation architectural heritage (Sana'a, Sa'ada) case study. In: Conference on technology and sustainability in urbanization. King Saud University, College of Architecture and Planning, Al-Riyadh, Saudi Arabia, pp 121–136
Al-sabahi HM (2005) A comparative analysis of the vernacular housing cluster of Yemen Sana'a and Shibam Hadhramaut a case study. J Sci Technol 10(1):27–34
Al-Saqqaf AMA (2002) The impact of natural and social environment on the properties of the mud architecture in Wadi Hadhramaut. Riyadh, Saudi Arabia
Avrami E, Guillaud H, Hardy M (2008) Terra literature review- an overview of research in earthen architecture conservation. The Getty Conservation Institute, Los Angeles
Azab KMM (1997) Planning and architecture of Islamic cities (Al-Ommah B). The Ministry of Awqaf and Islamic Affairs, Al-Doha
Baeissa AA (2014) Mud-brick high-rise buildings architectural linkages for thermal comfort in Hadhramaut Valley, Yemen. Int Trans J Eng Manag Appl Sci Technol 5(3):167–182
Boustead H (1960) The Hadhramaut. J R Cent Asian Soc 47(1):5–10
Brandstedt E (2013) The construction of a sustainable development in times of climate change. Lund University, Lund
Buchalter AR (2008) Country profile: YEMEN – library of congress. Federal Research Division, (August), 4–5. Retrieved from http://www.loc.gov/rr/frd/cs/profiles/Yemen.pdf
Bukair A, Al-Madhaji MS (2015) Sustainability and value of building and constructing residential buildings in Wadi Hadramaut – a case study of Shibam Hadramaut buildings. J Sci Technol 20(2):1–21
Carter JG, Cavan G, Connelly A, Guy S, Handley J, Kazmierczak A (2015) Climate change and the city: building capacity for urban adaptation. Prog Plan 95:1–66
Chandrasekar K, Krishnamurthy RR (2010) Chapter 10 climate change adaptation and coastal zone management. In: Shaw R, Pulhin JM, Pereira JJ (eds) Climate change adaptation and disaster risk reduction: issues and challenges, Community, environment and disaster risk management, vol 4. Emerald Group Publishing Limited, Bingley, pp 217–242
Coffman RL, (1990) Modification of the physical properties of natural and artificial adobe by chemical consolidation MRS proceedings 185
Costa CS, Rocha F, Velosa AL (2016) Sustainability in earthen heritage conservation. Geol Soc Lond, Spec Publ 416(1):91–100
Damluji SS (1986) The principles of architectural planning in Hadhramaut "Shibam & Terim": a case study the People's Republic of Yemen. Royal College of Art, London
Dethier J (1986) Des Architectures de Terre. Edition de Centre Pompidou, Paris
El Tahan AHMH, Elhanafy HEM (2016) Statistical analysis of morphometric and hydrologic parameters in arid regions, case study of Wadi Hadhramaut. Arab J Geosci 9(2):88–98
Farquharson FAK, Plinston DT, Sutcliffe JV (1996) Rainfall and runoff in Yemen. Hydrol Sci J 41(5):797–811
Hay J, Mimura N (2010) The changing nature of extreme weather and climate events: risks to sustainable development. Geomat, Nat Haz Risk 1(1):3–18
Ingrams WH (1935) House building in the Hadhramaut. Geogr J 85(4):370–372
Ingrams WH (1936) Hadhramaut: a journey to the Sei'ar Country and through the Wadi Maseila. Geogr J 88(6):524–551
Jerome P (2009a) Hadhramaut province, Yemen: report on the effects of the 23–24 October 2008 flash flood. New York
Jerome P (2009b) Shibam world heritage site Wadi Hadhramaut, Yemen: report on the effects of the 23–24 October 2008 flash flood. New York
Jerome P (2015) Earthen architecture: Yemeni Mudbrick at risk. Herit Risk, pp 263–264
Jerome P, Chiari G, Borelli C (1999) The architecture of mud: construction and repair technology in the Hadhramaut region of Yemen. APT Bull 30(2–3):39–48
Koellisch W (2015) Umm er Rasas, a world heritage site, mysterious and hidden. Preventive measures against damage from earthquakes and heavy rains. Herit Risk, pp 73–77.

Mehta D (2009) On conservation and development: the role of traditional mud brick firms in Southern Yemen. GLOBELICS 2009, 7th international conference, (2007), 1–23. Retrieved from http://hdl.handle.net/1853/35514

Muhlberger, S (1998) Early agriculture and early cities. Ontario, Canada

Nahal ALH (2010) Preserving architectural heritage within sustainable urban development: the case of Temasin Palace, south of Algeria. In: International conference on technology & sustainability in the built environment. Al-Riyadh, Saudi Arabia, pp 137–148

Noman AA (2005) Climate change in Yemen (No. 1). Al-Mukalla, Yemen

O'Brien G, O'Keefe P, Jayawickrama J, Jigyasu R (2015) Developing a model for building resilience to climate risks for cultural heritage. J Cult Herit Manag Sustain Dev 5(2):99–114

Pacheco-Torgal F, Jalali S (2012) Earth construction: lessons from the past for future eco-efficient construction. Constr Build Mater 29:512–519

Parry M, Canziani O, Palutikof J, van der Linden P, Hanson C (2007) Climate change 2007: impacts, adaptation, and vulnerability. Intergovernmental Panel on Climate Change, New York

Petzet M, Koenigs W (2015) Yemen – heritage at risk! Herit Risk, 195–198

Schipper ELF, Thomalla F, Vulturius G, Davis M, Johnson K (2016) Linking disaster risk reduction, climate change, and development. Int J Disaster Resilience Built Environ 7(2):216–228

Schroeder H (2016) Sustainable building with Earth. Springer, Cham

Shamshir F, Al-Mashhoor SH (2009) Architectural characteristics of Shibam City and the impact of the flood disaster. In: Second engineering conference. Aden University, Faculty of Engineering, Aden, pp 275–290

Sharma V, Marwaha BM, Vinayak HK (2016) Enhancing durability of adobe by natural reinforcement for propagating sustainable mud housing. Int J Sustain Built Environ 5(1):141–155

Stewart MG, Deng X (2015) Climate impact risks and climate adaptation engineering for built infrastructure. J Risk Uncertain Eng Syst 1(1):1–12

The Intergovernmental Panel on Climate Change (IPCC) (2012) Managing the risks of extreme events and disasters to advance climate change adaptation. IPCC

The Intergovernmental Panel on Climate Change (IPCC) (2014) Climate change 2014: synthesis report. The contribution of working groups I, II and III to the fifth assessment report of the intergovernmental panel on climate change. Retrieved from https://www.ipcc.ch/pdf/assessment-report/ar5/syr/SYR_AR5_FINAL_full_wcover.pdf

Van der Gun JAM, Ahmed AA, Saif A, Bashueib S, Negenman AJH (1995) The water resources of Yemen - a summary and digest of available information, report WRAY 35, GDHTTNO. Sana'a, Delft

Warner WHL (1931) Notes on the Hadhramaut. Geogr J 77(3):217–222

Wei H-H, Skibniewski MJ, Shohet IM, Yao X (2016) Lifecycle environmental performance of natural-hazard mitigation for buildings. J Perform Constr Facil 30(3):1–13

Wiebelt M, Breisinger C, Ecker O, Al-Riffai P, Robertson R, Thiele R (2011) Climate change and floods in Yemen. Development Strategy and Governance Division, Washington

Woodward M, Kapelan Z, Gouldby B (2014) Adaptive flood risk management under climate change uncertainty using real options and optimization. Risk Anal 34(1):75–92

Chapter 4
On the Economics of Climate Change and Health: An Overview

Popescu George Cristian and Ion Andreea Raluca

Abstract The World Health Organization acknowledges several key directions in which climate change impacts on health. Extreme heat is responsible for many deaths from cardiovascular and respiratory diseases and contributes significantly to spreading allergens. The prevalence of malnutrition and undernutrition increases with the floods and heavy rainfalls, while the pattern of infections is also influenced, and the incidence of cancers is on the rise. In the same time, the research community admits that isolating and measuring the effects of climate change on health is a very difficult task. A lot of work has been carried recently for this purpose, revealing many promising directions for further exploration. The aim of this chapter is two-fold. On one side, it intends to review the existing literature in the area of health and climate change, with a focus on the methods and methodologies used until now. On the other side, we provide a discussion on the specificities of the Central and Eastern European case, emphasizing the particular role of perception and subjective beliefs on the topic of climate change and global warming in this region. The main findings show that climate change has a negative impact on human health, in two main directions: air pollution and increasing temperatures. Pollution affects negatively human health; the results of the experimental sections demonstrating that there is a high correlation between the levels of CO_2 emissions, as an indicator of pollution, and the incidence of cancer show the health status. Increasing temperatures negatively affect human health, the most affected areas in Europe being Central and Eastern European countries. In this region, an increase of 0–5 °C of the average temperatures, due to climate change, leads to illnesses caused by microorganisms, which survive and multiply in a warmer environment.

Keywords Climate change · Health · Air pollution · Human Development Index

---

P. G. Cristian (✉)
Faculty of Business and Administration, The University of Bucharest, Bucharest, Romania
e-mail: cristian.popescu@faa.unibuc.ro

I. A. Raluca
Faculty of Agrofood and Environmental Economics, The Bucharest University of Economic, Bucharest, Romania

A. Omran, O. Schwarz-Herion (eds.), *The Impact of Climate Change on Our Life*, https://doi.org/10.1007/978-981-10-7748-7_4

4.1 Baseline Relationships Between Climate Change and Health

Health is a general concern of all people. Health status depends on the continued stability and functioning of the biosphere's ecological and physical systems (WHO 2003a, p.1). In the last decades, the life support system, including the climate system, worsened and distanced from the natural system, because of high pressure of economic activities and industrialization process. As a result, people's health worsened, and more illnesses appeared. At the beginning of the twenty-first century, the relationships between climate change and health are complex. There is near unanimous scientific consensus that climate change has a negative impact on human health, due to higher temperatures and changes in precipitations, worsened air quality, more intense severe weather, and rising sea levels. All these effects harm human health and well-being. Although people have acclimatized and adapted to local climates and also are able to cope with a range of weather changes, there is a range of individual sensitivity to extreme weather events (WHO 2003a, p.47). The main climate drivers identified in literature (USGCRP 2016) to which people must adapt are:

- Extreme heat, due to more frequent, severe, prolonged heat events
- Outdoor air quality, due to increasing temperatures and changing precipitation patterns
- Flooding, due to a rising sea level and more frequent or intense extreme precipitations, hurricanes, and storm surge events
- Vector-borne infection, due to changes in temperature extremes and seasonal weather patterns
- Water-related infection, due to rising sea surface temperature, changes in precipitation, and runoff affecting coastal salinity
- Food-related infection, due to increases in temperature, humidity, and season length
- Mental health and well-being, due to climate change impacts especially extreme weather.

Synthesizing the drivers above, they can be classified into two categories, depending on their direct and indirect effects on health (WHO 2003a, p.47–51).

4.2 The Direct Effect of Climate Change on Health

Heatwaves and other extreme events are the most recognized climate change drivers to health challenges. Extreme weather events, including periods of very high temperature, torrential rains and flooding, droughts, and storms, stress populations beyond their adaptation limits (McMichael et al. 2006). Mortality rates rise at temperatures outside the comfort zone (Curriero et al. 2002). But temperature extremes have also beneficial effects, as found by McMichael et al. (2006) because they lead to reduced winter deaths and disease events in some temperate countries. Studies (WHO 2003a, p.48) show that climate-related disasters have caused hundreds of thousands of deaths in countries such as China, Bangladesh, Venezuela, and Mozambique.

Outdoor air quality is affected by climate change and has negative implications on human health. Heat waves and decreasing precipitations lead to increases in the ozone and particulate matter, driving to cardiovascular and respiratory illnesses and death (WHO 2003b). Climate change influences the transportation of airborne pollutants, pollen production, and levels of fossil fuel pollutants resulting from household heating and energy demands (WHO 2003a, p.48). Moreover, experimental studies (Ziska and Caulfield 2000) found that doubling CO_2 levels induces a fourfold increase in the production of ragweed pollen, causing allergen diseases. Other studies (Beggs 2004) underline the same ideas that climate change might also alter the timing and duration of pollen and spore seasons and the geographic range of these aeroallergens, affecting allergic disorders such as hay fever and asthma.

Floods result from the interaction of rainfall, surface runoff, evaporation, wind, sea level, and local topography (McMichael et al. 2006). They have complex effects on human health: drowning, injuries, mental health consequences, gastrointestinal and other illness (National Institutes of Health 2016), or exposure to toxic pollutants (Stachel et al. 2004). Floods have recently tended to intensify because of climate change (Milly et al. 2002).

4.3 The Indirect Effects of Climate Change on Health

Infectious diseases include malaria, dengue fever, and yellow fever (all mosquito-borne), various types of viral encephalitis, schistosomiasis (water snails), leishmaniasis (sand flies: South America and Mediterranean coast), Lyme disease (ticks), and onchocerciasis (West African river blindness, spread by black flies) (WHO 2003a). These diseases are caused by many factors, including social, economic, climatic, and ecological conditions (Weiss and McMichael 2004). Infectious diseases can be vector-borne, water-related, or food-related infections.

(i) *Vector-borne infectious diseases*. Due to increasing temperatures and changing seasonal weather patterns, the range and seasonality of vectors will be extended. The ticks show earlier seasonal activity, increasing the risk of human exposure to Lyme disease-causing bacteria (National Institutes of Health 2016); they enlarge their geographic areas to North in Sweden (Lindgren et al. 2000) and to altitude in the Czech Republic (Danielova 1975). Reports (WHO 2003a, p.49) show that in areas with limited or deteriorating public health infrastructure, increasing temperatures expand the geographic range of malaria transmission to higher altitudes and latitudes. The most vector-borne infectious disease-affected areas are tropics and subtropics.

(ii) *Waterborne infectious diseases*. Human health is influenced by water quality, sanitation, and hygiene. Rainfall events can transport terrestrial microbiological agents into drinking water sources resulting in outbreaks of cryptosporidiosis, giardiasis, amoebiasis, typhoid, and other infections (WHO 2003a, p.49). Increases in water temperatures lead to changes in the marine environment that alter risks of biotoxin poisoning from human consumption of fish and shellfish and alter timing and location of *Vibrio vulnificus* growth, raising the exposure and risk of waterborne illness (National Institutes of Health 2016).

(iii) *Food-related infectious diseases.* Rising temperatures and humidity increase pathogens and *Salmonella* prevalence in food. Longer seasons and warming winters increase the risk of exposure and infection.

Mental health and well-being. Population displacement resulting from sea level rise, natural disasters, or environmental degradation is likely to lead to substantial health problems, both physical and mental (WHO 2003a, p.49).

Beyond these specific climate change risks to health, there are indirect and more complex health effects due to the social, economic, and political disruptions of climate change, including effects on regional food yields and water supplies (McMichael et al. 2006).

4.4 Climate Change and Health in Action

This section presents a regional and structural overview of the relationships between weather and health, and it describes the human health consequences of climate in different regions, including Europe and Romania, and by categories, as they are divided in the report of the Interagency Working Group on Climate Change and Health (Portier et al. 2010), outlining the research needs on the human health effects of climate change.

Asthma, respiratory allergies, and airway diseases may become more prevalent because of increased human exposure to pollen (due to altered growing seasons), molds (from extreme or more frequent precipitation), air pollution and aerosolized marine toxins (due to increased temperature, coastal runoff, and humidity), and dust (from droughts) (Portier et al. 2010). According to the Eurostat Statistics Database, self-reported asthma in Romania is 2% of the population, meaning 399,668 persons (EUROSTAT 2017), while Western and Northern countries report higher levels: 9.4% in the United Kingdom, 9.2% in Finland, 9% in Ireland, 8.9% in Iceland, and 8.8% in France. According to unofficial data, every fourth Romanian citizen suffers from a certain type of allergy (Statistics of respiratory tract diseases. Romania). Experts (D'Amato and Cecchi 2008) speculate that the global rise in asthma was indirectly related to climate change.

Cancer may become more prevalent because of increased duration and intensity of ultraviolet (UV) radiation and because of human exposure to chemicals and toxins in a polluted environment. In Romania, the incidence of malignant neoplasms is high, basically the second cause of death after diseases of the circulatory system, killing more than 50,000 persons every year (Eurostat 2017). Lung cancers, with about 10,000 deaths (in 2014) account for about 20% of overall deaths from cancer in Romania. The main cause of lung cancer is smoking, but air pollution (Beelen et al. 2008), including indoor air pollution, and fine particulates also contribute to the burden of lung cancers. One of the possible impacts of climate change on cancer is an increased ultraviolet (UV) radiation exposure which increases the risk of skin cancers and cataracts (Tucker 2009).

Cardiovascular diseases and stroke intensify by increasing heat stress, increasing the body burden of airborne particulates, and changing the distribution of zoonotic vectors that cause infectious diseases linked with cardiovascular disease (Portier et al. 2010). Diseases of the circulatory system are the leading cause of death in Romania, with 149,070 deaths in 2014. In 2015, 573,663 people (almost 3% of total population) were suffering of diseases of circulatory system, as statistics of people discharged from hospitals show (Eurostat 2017). Weather conditions, both extreme cold and extreme heat, serve as stressors in individuals with pre-existing cardiovascular disease. Climate change is also implicated indirectly in cardiovascular disease, as shown by specialists (Portier et al. 2010): the incidence of certain vector-borne and zoonotic diseases (VBZD) with cardiovascular manifestations.

Foodborne diseases and nutrition problems appear as a result of changes in climate which are associated with staple food shortages, malnutrition, and food contamination (of seafood from chemical contaminants, biotoxins, and pathogenic microbes and of crops by pesticides). Instead of being a source of essential nutrients, in the context of extreme heat, food can be a source of exposure for foodborne illness, resulting from ingesting food that is spoiled or contaminated with microbes and chemical residues such as pesticides, biotoxins, or other toxic substances. In Romania, 619 new cases of food poisoning were reported in 2013, down from 1781 new cases in 2008 (National Institute of Statistics 2014, p.278). Foodborne illnesses include not only food poisoning but also campylobacteriosis and salmonellosis infections. In Romania, 1539 new cases of salmonellosis were reported in 2015. Drops in food production levels result as a consequence of extreme weather events such as droughts, flooding, and hurricanes. Not only climate change but also agricultural practices are responsible for chemical residues in agricultural products, because of their intensive forms of growing both crops and animals and excessive use of fertilizers, pesticides, antibiotics, and growth stimulators.

Heat-related morbidity and mortality increase in response to climate change. The health outcomes of prolonged heat exposure include heat exhaustion, heat cramps, heat stroke, and even death (Ellis 1976). Additional illnesses can result from prolonged exposure to heat by exacerbating pre-existing chronic conditions such as various respiratory, cerebral, and cardiovascular diseases (Kovats and Hajat 2008). The World Health Organization (Campbell-Lendrum and Woodruff 2007) estimated 166,000 deaths, and about 5.5 million disability-adjusted life years (DALYs, a measure of overall disease burden) were attributable to climate change in 2000, worldwide. The same report shows that climate change effects on malnutrition, diarrhea, and vector-borne diseases appeared to be considerably more important than effects on flooding or on deaths attributable to thermal extremes. Studies in Europe (Gosling et al. 2009) reported deaths caused by sensitivity to cerebrovascular disease and sensitivity to cardiovascular disease-related mortality associated with heat had been seen especially among the elderly. The same study reported for Romania 26 deaths caused by extreme heat in June to July 2006, 13 deaths in June to August 2005, and 27 deaths in July 2004. The highest numbers of deaths caused by extreme heat in August 2003 have been reported in Italy, 20,098 deaths; France, 19,490 deaths; and the United Kingdom, 2,045 deaths.

Human development may be affected by climate change in two ways. One is malnutrition, particularly during the prenatal period and early childhood as a result of decreased food supplies. Specialists (Portier et al. 2010) showed that maternal undernutrition acts on the developing fetus in ways that program the risks for adverse health outcomes such as cardiovascular disease, obesity, and metabolic syndrome in adult life. Fetus exposure to certain chemicals or biotoxins due to climate change impacts the health of future adults, which, afterward, lead to increasing societal costs in terms of medical care. The other way in which the human development may be affected by climate change is the exposure to toxic contaminants and biotoxins, resulting from extreme weather events and increased pesticide use for food production, and increases in harmful algal blooms in recreational areas (Portier et al. 2010).

Mental health- and stress-related disorders may occur in cases in which climate change effects imply the geographic displacement of populations, damage to property, loss of loved ones, and chronic stress (Fritze et al. 2008). All these can negatively affect mental health and manifest themselves through acute or prolonged anxiety, obsessive-compulsive disorder, and post-traumatic stress disorder. Studies carried out in Europe (WHO 2015) reported that across the European Region, neuropsychiatric disorders are the second largest contributor to the burden of disease, accounting for 19% of the total. About 1–2% of the population is diagnosed with psychotic disorders, men and women equally, and 5.6% of men and 1.3% of women have substance use disorders. Extreme weather events such as hurricanes, wildfires, and flooding may lead to increased anxiety and emotional stress about the future.

Neurological diseases and disorders may appear as a result of human exposure to biotoxins (from harmful algal blooms), metals (found in new battery technologies and compact fluorescent lights), and pesticides (used in response to changes in agriculture) (Portier et al. 2010). Moreover, it has recently been reported that exposure to pesticides and herbicides may increase the risk of neurological diseases (Costello et al. 2009). The health effects identified in the literature (Wang 2008) include amnesia, diarrhea, numbness, liver damage, skin and eye irritation, respiratory paralysis, and even death.

Vector-borne and zoonotic diseases may increase as a result of climate change due to related expansions in vector ranges, shortening of pathogen incubation periods, and disruption and relocation of large human populations. In Europe, tick-borne encephalitis (TBE) and Lyme borreliosis (Lyme disease) are the two most important tick-borne diseases, both of which are transmitted primarily by *Ixodes ricinus*. Lyme disease is the most common vector-borne disease in the EU, with a reported incidence of approximately 65,000 cases per year (Medlock et al. 2013).

Waterborne diseases appear as a result of water contamination with harmful pathogens and chemicals due to increases in water temperature, precipitation frequency and severity, evaporation-transpiration rates, and changes in coastal ecosystem health (Portier et al. 2010). Waterborne microorganisms, as described by Batterman et al. (2009), include protozoa that cause cryptosporidiosis, parasites that cause schistosomiasis, bacteria that cause cholera and legionellosis, viruses that cause viral gastroenteritis, amoebas that cause amoebic meningoencephalitis, and algae that cause neurotoxicity. The majority of waterborne disease is gastrointestinal, though they can manifest themselves as respiratory infections as well or as other illnesses with immunologic, neurologic, hematologic, metabolic, pulmonary, ocular, renal, and nutritional complications (Meinhardt 2006).

Weather-related morbidity and mortality increase as a result of increases in the incidence and intensity of extreme weather events such as hurricanes, floods, droughts, and wildfires. The human health consequences of climate change described before are interconnected and intersect each other. For example, extreme weather events lead to waterborne and vector-borne illnesses following floods and hurricane storm surges, as well as malnutrition resulting from losing land and animals and damaging agricultural production. Moreover, mental health- and stress-related disorders may occur as a result of personal tragedy and displacement.

4.5 Research Dircctions

The appropriate task of researchers studying climate changes is to make estimations of the consequent health effects of climate change, with the final goal of preventing diseases. The main questions that arise are "How will the climate system respond to future higher levels of greenhouse gases?" and "How will human health be affected by future climate changes?"

The relevance of this research results from the need to expand the understanding of the often indirect, long-term, and complex consequences of climate change for human health. The research starts from the premises that climate change has a negative impact on human health. The negative effects are more significant for older age groups and people living in urban areas, who are at greater risk to illness caused by climate change than young people and those in nonurban areas. This hypothesis is consistent with other findings showing that people living in urban environments are at greater risk than those in nonurban regions (Smoyer et al. 2000). Also, McGeehin and Mirabelli (2001) show that urban heat island effect, consisting in high thermal mass and low ventilation, where the environment absorbs and retains heat, amplifies and extends the rise in temperatures, especially overnight. Moreover, if heat waves increase, the risk of death and serious illness would increase principally in the urban poor, older age groups, and those with pre-existing cardiorespiratory diseases (WHO 2003a, p.47).

Among the climate change drivers described before, we focus our research on air quality, because it represents the cause of the determinants of climate change.

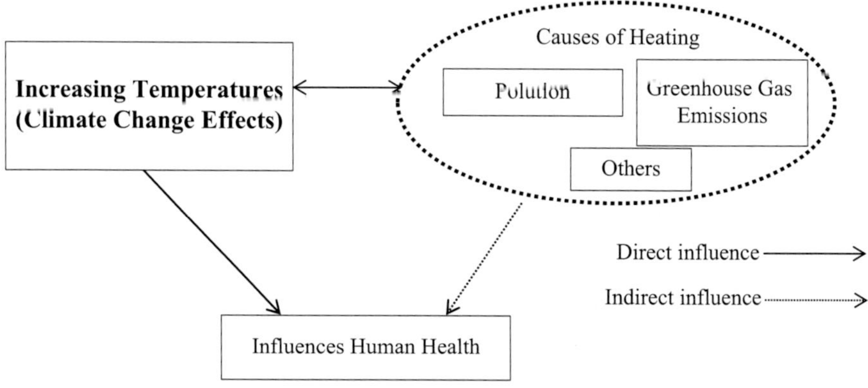

Fig. 4.1 The impact of temperature increases on human health

Greenhouse gas emissions generated by human activity are responsible for climate change, and this is the reason why we consider necessary to assess the relationships between air pollution and human health. In pursuing this, one hypothesis that we intend to test in this chapter is that air pollution negatively influences human health (*H1*).

WHO reports (2003) outline the main tasks for researchers studying climate change and health. Among them, the sensitivity of particular health outcomes to weather, climate variability, and climate-induced environmental changes needs to be studied (WHO 2003b, p.13). Our research focuses on this issue, trying to validate the hypothesis *H2* (increasing temperature negatively influences human health). We start our research from the assumption that the major pathogens multiply faster in warmer conditions.

The impact of temperature increases on human health has both direct and indirect consequences (Fig. 4.1). We argue that temperatures indirectly influence people's health when they indirectly interrelate between the cause-effect relationships determined by the rise in temperatures – the impact on human health. In this relationship, greenhouse gas emissions would imply increasing temperatures but also pollution in all its forms (air, soil, water, food). Increasing temperature and pollution are the two causes whose phenomena mutually condition themselves with an indirect effect on the degradation of human health, as illustrated in Fig. 4.1.

4.6 Experimental Section

The effects of pollution on people's health are reported in different specialized writings. They are outlined differently for children and adults. The first target group is more vulnerable in the first year of life, as found in studies showing the impact of air pollution among the primary causes of illnesses for children, especially up to 1 year (Gehring et al. 2002). In validating the hypothesis *H1* (air pollution negatively influences human health), we start our research from comparing and correlating the data regarding the air pollution (in CO_2 emissions) with a health indicator, such as the Human Development Index (HDI). This indicator is taken into account by the WHO in the reports that justify the incidence of cancer among the population (as the main disease that spread along with the development of world economies – World Cancer Report 2014). In the literature, there are other indicators, such as Functional Status Index, FSI (Fanshel and Bush 1970), which is the result of several indicators, *HDI* Human Development Index, *HPI* Happy Planet Index, *GPI* Global Peace Index, *DI* Democracy Index, and *II* Income Inequality. The Functional State Index is considered one of the most sensitive indicators linking human health with the intensive development process based on the polluting industry that leads to environmental degradation and ultimately to health degradation (Blaxter 2003). However, due to the complexity of HDI that takes into account life expectancy, education, and the decent standard of living, we consider it an indicator that can be correlated with the degree of pollution and the evolution of cancerous diseases over a period of time.

One methodological problem is that pollution usually has delayed effects on the evolution of health indicators, which has led us to "fade" the comparison of data series according to medical studies. Thus, any environmental report shows that the effects of air pollution on human health (Annual report – The state of the environmental factors in Romania 2010) can appear after years, even decades, and the most vulnerable population being the child population and the groups over 65 years of age. The links between pollutant emissions and the health status are made taking into account a long period of time for two indicators. The first one is that showing the degree of global pollution, such as the pollution resulting from the combustion of fossil fuels through CO_2 emissions (the other substances harmful to health being associated with this indicator). The second one is the incidence of cancer, showing the health status. In view of the above, we compare the average CO_2 emissions per capita from 1995 to 2009 in developed countries and the health indicator: the incidence of cancer as an average over the period 2003–2009.

From all considerations above, the proposed indicators for lifestyle measurement with impact on human health as a result of environmental degradation would be:

- Public health indicators: healthy years of life, life expectancy at birth by gender, indicators of illness, and death for diseases influenced by lifestyle
- Other proposed indicators: Human Development Index (HDI), calculated as the geometric mean of indicators of healthy life expectancy, education, and the decent standard of living

In analyzing data on the incidence of cancer, the rate at 100,000 thousand inhabitants, as an average of the period 2003–2007, according to the World Cancer Report 2014, we will make the first adjustment of the reporting at a rate of 10,000 inhabitants to be representative of this indicator on the same graph with the atmospheric pollution indicator per unit of CO_2 per capita reported over the period for the same countries. The calculations on the correlation of these data series have shown that there is a high correlation, the coefficient of correlation being $R = 0.70$. When comparing the HDI with the pollutant emissions, the correlation of the data is closer, the

Table 4.1 The top countries with the highest incidence of cancers, compared to the HDI level and the CO_2 emissions

Country	Cancer incidence (rate to 10,000 inhabitants, as average of the period 2003–2007)	HDI (Human Development Index), very high	CO_2 emissions/inhabitant (average 1995–2009)
United States of America	34.7525	0.914	16.9
Australia	32.585	0.933	18
Denmark	30.855	0.9	7.9
Slovakia	27.645	0.83	8.1
Spain	26.125	0.869	6.3
Japan	21.805	0.89	9.2

Source: Author calculations based on statistical data

coefficient being 0.75. Considering the components of this Human Development Index, we may argue that pollution takes into account the level of education and living standards, important components in the use of natural resources, and their capitalization. Centralizing the data computed, the results presented in Table 4.1 illustrate the top countries with the highest incidence of cancers to 10,000 inhabitants, as an average of the period 2003–2007, compared to the HDI level and the CO_2 emissions in the same countries.

At the opposite end, data on cancers and mortality in children (under 15 years) as ration of the total population as an average for the period 2005–2012 are centralized (Fig. 4.2). Data show that there is an inverse correlation between HDI and the proportion of cancers in children, and the developing countries have a high rate of cancers in children, and developed countries have a low proportion of cancers in children.

Not the same thing can be argued for adults when assessing the effects of pollution on human health over a longer period of time. Thus, in developed countries, which register high levels of the HDI indicator, long-term CO_2 emissions have registered high levels as well. Moreover, there is also a high incidence of cancers over periods of time comparable from the point of view of the harmful effects of pollution on human health. In Fig. 4.3, the indicators are centralized and presented in Table 4.1 with the centralizing data; this close correlation between data is observed.

All these data show that the population's health is closely related to the pollutants and *H1* (air pollution negatively influences human health) is validated, according to the results presented in Table 4.1 and Figs. 4.1 and 4.2. This conclusion is all the more important because it does not know yet the precise effects of different substances on the evolution of cancerous diseases (their causes), but the statistical and econometric research (Chirila et al. 2012) demonstrated that the health degradation is caused by the lack of adaptation of the lifestyle to the environmental conservation component. Thus, there are different categories of people who, due to their unadjusted lifestyle, are more exposed to cancer. We argue that there are one or more drivers, according to the literature, related to adapting lifestyle to human health and can result in a greater proportion of illness of an organ affected by that

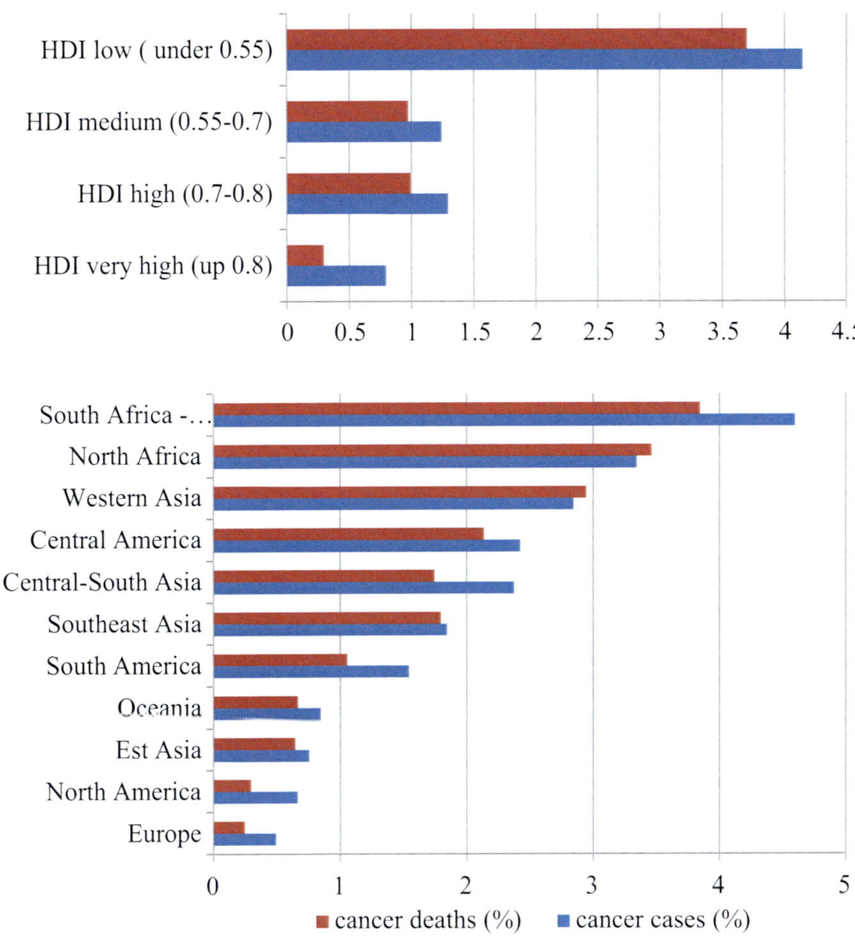

Fig. 4.2 Correlations between HDI and the proportion of cancers in children (Source: Author calculations based on statistical data)

pollutant (Table 4.2). Studies (WCR) show that the best example is given by the close link between cancer, over the last period of time, and cancer mortality, which has ranked second from fifth place, in just 10 years. Moreover, forecasts show that in the following 10 years, it will rank first in the world mortality index (according to WCR estimates 2014). We reported this disease because on the one hand the development of cancer is closely related to increased pollution, as we have shown, while on the other hand, in the literature there are studies showing the effect of pollution on humans in all respects: water, air, soil, food pollution, and electromagnetic pollution (as the main sources influencing people's lives and health) – according to the International Agency for Research on Cancer.

As seen in Table 4.2, the lifestyle has a significant impact on health, leading to cumulated risks of 70%, considering the fact that humans live in a natural environ-

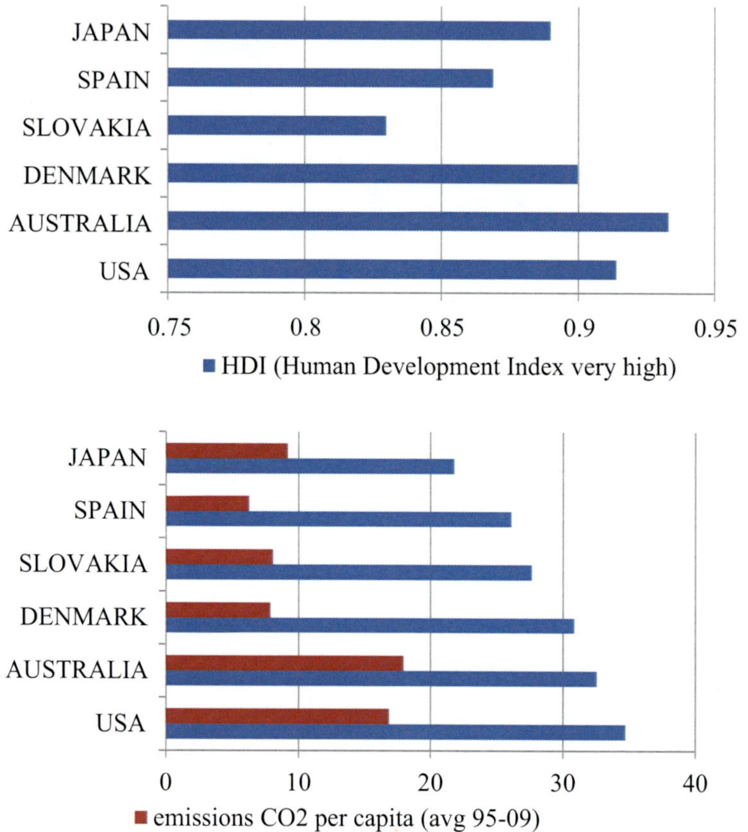

Fig. 4.3 Correlations between HDI, the incidence of cancers in adults and the CO_2 emissions per capita (the most important countries with very high HDI but with high level of CO_2 emissions and very high incidence of cancer rates as average on period 2000–2007) (Source: Author calculations based on statistical data)

Table 4.2 Cancer risk factors

Factor	Impact (%)
Food deficiency	30
Smoking	30
Hereditary factors	15
Obesity, sedentary lifestyle	5
Professional risks	5
Infections	5
Alcohol	3
Drugs	2
Pollution	2
UV exposure	2

Source: Beliveau and Gingras (2006)

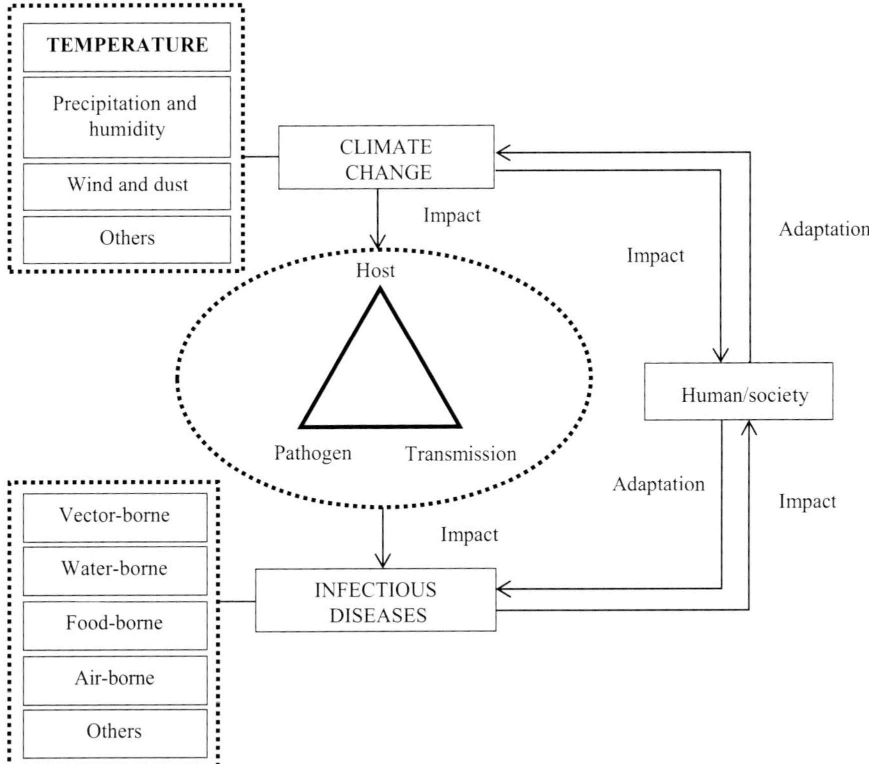

Fig. 4.4 Climate change, human infectious diseases, and human society (Source: Xiaoxu Wu et al. 2016)

ment in which they are born, develop themselves, and then they die. In all this cycle, people's life is influenced by aggressive external factors considered pollutants for human health. Coming back to Fig. 4.1 (illustrating the impact of temperature increases on human health) and to *H2* (increasing temperature negatively influences human health), hereinafter, the impact of increasing temperatures on human health is assessed. This approach is a topic as presently controversial, because the assessment itself requires knowledge from vast areas of human health, environmental, economics, and other scientific disciplines. There are many specialists who have shown this direct causal link between rising temperatures and environmental protection in the overall climate change (Wu et al. 2016). Thus, there is a whole series of diseases that can occur with greater probability due to the occurrence and multiplication of viruses, bacteria, or other organisms or insects that can withstand the limits of frost temperatures (Fig. 4.4), or which do not develop anymore, or do not multiply at low temperatures (food germs). Thereby, there is a whole range of bacteria, viruses, but also microbial insects, for which an increase in temperatures of 2 °C can make the difference between freezing and thawing, or otherwise, between life and the survival and multiplication of these pathogens (Schvoerer et al. 2008).

The logical scheme through which climate change affects people's health is illustrated in Fig. 4.4, in which the author shows that there are a multitude of factors that are included in climate change and affect people's health, including changes in temperatures, in particular, their growth. From the abovementioned, the temperature increases by a few degrees as annual averages will affect those regions that, due to the increase of the minimum temperature registered during one year, will lead to the survival and creation of atmospheric conditions favoring the survival and multiplication of pathogenic germs. Thus, the minimum temperatures at which most *Bacillus* species breed are about 5 °C (Moldoveanu 2012). *Clostridium botulinum-bacteria*, types A, B, C, and D are no longer multiplying below 5.6 °C and no longer produce toxins under these temperatures (Moldoveanu 2012). *Staphylococci*, whose presence in the human body can cause various diseases (*aureus*, causing among others, skin infections), are found in foods stored at temperatures above 10 °C, and *enterococci* survive and can breed at temperatures above 10 °C (Jackson et al. 2007). For example, *enterococci* can survive and multiply in milk and also in water, which constitute a survival and multiplication support, to the minimum temperature of 10 °C (Jakson et al. 2007). *Salmonella alte Enterobacteriaceae*, being species of bacteria derived from *Salmonella*, have minimum breeding temperatures between 0 and 10 °C. Therefore, this bacterium does not proliferate below the freezing rate. Thus, there are a whole series of bacteria and viruses with known negative effects on human health, which at constant low temperatures below 5 °C inhibit their propagation functions and which no longer develop or survive at freezing temperatures below 0 °C. Thus, a significant threshold for the increase in annual average temperatures is 0 °C, in the coldest months of the year. To exemplify the effects of global warming on human health, the impact of temperature increases in Europe is considered in the following. Thus, according to Eurostat, January is known as the month with the lowest average annual temperatures. If we plan to represent the temperature increases by several degrees C, we can notice, in Fig. 4.5 – the area in Europe affected by the rise of the annual average temperatures (up to 2–5 °C), with an immediate impact on human health, is Eastern Europe, and it is the most affected area. In this region, there are countries with the annual minimum temperature range around 0 °C.

According to Fig. 4.5, the area affected by the rise of temperature by only 2–5 °C is the climatic zone in Europe with continental temperatures, where, in winter, the average annual temperatures reach 0–5 °C. This range could increase to over 0–5 °C, due to climate change, respectively; it could be recorded as average annual positive temperatures. This would generate a whole series of issues with an impact on human health by creating an environment conducive to the survival and development of pathogens that favor illness. The regions affected by global warming of just 2–5 °C, relative to the Europe continent, are the area of Eastern and Central Europe. This region is relatively large, accounting for one third of the entire continental surface, especially the plains and plateaus. The affected area of Eastern and Central Europe cover about 1.5 million square km, from territories belonging to the countries Bulgaria, Romania, Ukraine, Moldova, Belarus, the Baltic States, Poland, Serbia, Macedonia, Bosnia, Austria, Hungary, the Czech Republic, Slovakia, Slovenia,

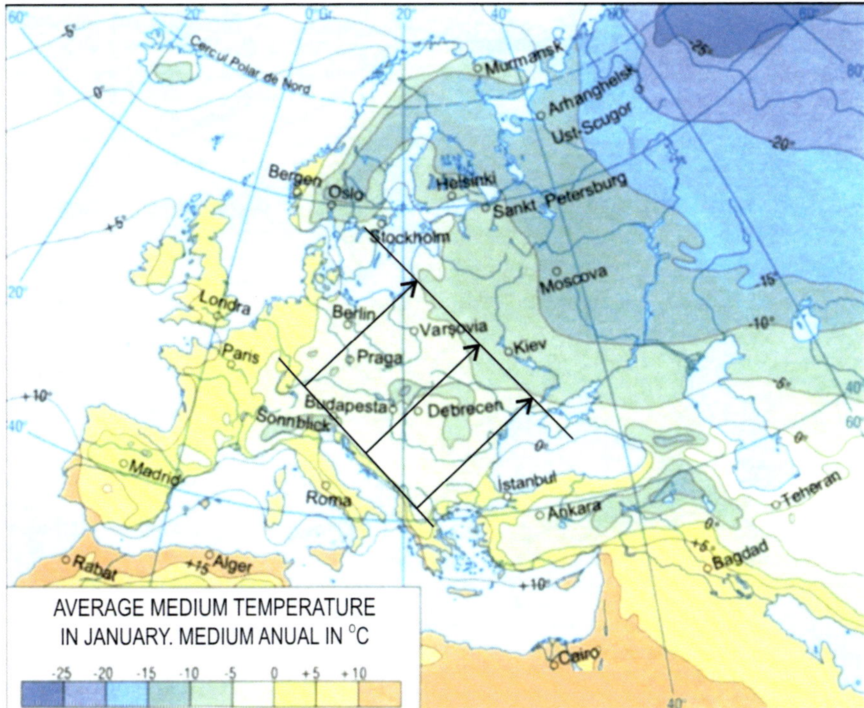

Fig. 4.5 The area (moving impact) in Europe affected by the rise of the annual average temperatures (up to 2–5 °C), with immediate impact on human health (Source: Geopolitical Maps – GIS)

Austria, and Germany. The approximate population in this area is over 170 million inhabitants, who will be affected by the increasing temperatures generated by climate change. Thus, the hypothesis "increasing temperature negatively influences human health" (*H2*) is validated, in Europe.

4.7 Conclusions and Recommendations for Action

The research aimed at quantifying the combined effects of exposure to heat waves and ambient air pollution on excessive illness and death. The main findings show that climate change has negative effects on human health, in at least two directions: through air pollution and through increasing temperatures. Pollution affects negatively human health; the results of the experimental sections show that there is a high correlation between the levels of CO_2 emissions, as an indicator of pollution, and the incidence of cancer, showing the health status.

Increasing temperatures negatively affect human health, the most affected areas in Europe being Central and Eastern European countries. In this region, a 0–5 °C increase of the average temperatures, due to climate change, leads to illnesses caused by microorganisms. Many bacteria, viruses, but also microbial insects will survive and multiply in an increasing climate, in which a temperature growth of 2 °C can make the difference between life and death for these pathogens.

The climate change negative effects can be mitigated through actions taken to reduce greenhouse gas emissions and to enhance the sinks that trap or remove carbon from the atmosphere to reduce the extent of global climate change (Portier et al. 2010). The major targets of climate change mitigation strategies include alternative fuels and energy conservation, changes in land use patterns, sustainable development of the built environment, and carbon capture and storage (Haines et al. 2006). The first one refers to changing the fuels from fossils ones to green energy, which will reduce the emissions of CO_2 into the atmosphere. Changes in land use refer to reducing the destruction of forests and reforestation some areas, considering the fact that about 80% of CO_2 is caused by industrialization and the rest by land use such as deforestation. Research shows future trends of climate change and health worldwide, including Europe. According to Sterl et al. (2008), by 2100 northeast India and Australia can expect summer temperatures to peak over 50 °C, and the southwest, central west, and southern Europe over 40 °C (Ulbrich et al. 2008) forecast increasing storm-track activity in the eastern North Atlantic, Western Europe, and parts of Asia, which might intensify extreme cyclones. Rockel and Woth (2007) estimate a future increase of up to 20% in the number of storm peak events over central Europe. In this context, the issue of climate change effects on human health becomes more important.

References

Annual Report – The State of Environmental Factors in Romania, INSSE, 2010

Batterman S, Eisenberg J, Hardin R, Kruk ME, Lemos MC, Michalak AM, Mukherjee B, Renne E, Stein H, Watkins C, Wilson ML (2009) Sustainable control of water-related infectious diseases: a review and proposal for interdisciplinary health-based systems research. Environ Health Perspect 117(7):1023–1032

Beelen R, Hoek G, van den Brandt PA, Goldbohm RA, Fischer P, Schouten LJ, Armstrong B, Brunekreef B (2008) Long-term exposure to traffic-related air pollution and lung cancer risk. Epidemiology 19(5):702–710

Beggs P (2004) Impact of climate change on aeroallergens: past and future. Clin Exp Allergy 34:1507–1513

Beliveau F, Gingras D (2006) L'alimentazione anti-cancro. (ed) Sperling & Kufer, Segrate, p 78

Blaxter M (2003) Health and lifestyle. TR Pb, Fallbrook, p 102

Campbell-Lendrum D, Woodruff R (2007) In: Prüss-Üstün A, Corvalán C (eds) Climate change: quantifying the health impact at national and local levels, WHO environmental burden of disease series No. 14. World Health Organization, Geneva

Chirila P, Popescu EM, Georgescu C (2012) Nu Hrani Cancerul. (ed) Christiana, Bucharest, pp 8–12

Costello S, Cockburn M, Bronstein J, Zhang X, Ritz B (2009) Parkinson's disease and residential exposure to maneb and paraquat from agricultural applications in the central valley of California. Am J Epidemiol 169(8):919–926

Curriero FC, Heiner KS, Samet JM, Zeger SL, Strug L, Patz JA (2002) Temperature and mortality in 11 cities of the eastern United States. Am J Epidemiol 155(1):80–87
D'Amato G, Cecchi L (2008) Effects of climate change on environmental factors in respiratory allergic diseases. Clin Exp Allergy 38(8):1264–1274
Danielova V (1975) Overwintering of mosquito-borne viruses. Med Biol 53:282–287
Ellis FP (1976) Heat illness. I. Epidemiology. Trans R Soc Trop Med Hyg 70(5-6):402–411
EUROSTAT (2017) European Commission, DG Health & Consumers, Public health. Available online at: http://ec.europa.eu/health/dyna/echi/datatool/index.cfm?indlist=26a. Accessed in June 2017
Fanshel S, Bush JW (1970) A health-status index and its application to health-services outcomes. Oper Res 18(6):1021–1066
Fritze JG, Blashki GA, Burke S, Wiseman J (2008) Hope, despair and transformation: climate change and the promotion of mental health and wellbeing. Int J Ment Heal Syst 2(1):13. https://doi.org/10.1186/1752-4458-2-13.
Gehring U, Cyrys J, Sedlmeir G, Brunekreefz B, Bellander T, Fischer P, Bauer CP, Reinhardt D, Wichmann E, Heinrich J (2002) Traffic-related air pollution and respiratory health during the first 2 years of life. Eur Respir J 19:690–698
Geopolitical Maps – GIS. http://www.geotutorials.ro/atlas-geografic/harti-europa/atlas-europa/harta-temperaturii-medii-anuale-luna-ianuarie-harta-europa/
Gosling SN, Lowe J, McGregor G, Pelling M, Malamud BD (2009) Associations between elevated atmospheric temperature and human mortality: a critical review of the literature. Clim Chang 92(3-4):299–341
Haines A, Kovats RS, Campbell-Lendrum D, Corvalan C (2006) Climate change and human health: impacts, vulnerability, and mitigation. Lancet 367(9528):2101–2109
Jackson V, Blair IS, McDowell DA, Kennedy J, Bolton DJ (2007) The incidence of significant foodborne pathogens in domestic refrigerators. Food Control 18:346–351
Kovats RS, Hajat S (2008) Heat stress and public health: a critical review. Annu Rev Public Health 29:41–55
Lindgren E, Talleklint L, Polfeldt T (2000) Impact of climatic change on the northern latitude limit and population density of the disease transmitting European tick Ixodes ricinus. Environ Health Perspect 108:119–123
McGeehin MA, Mirabelli M (2001) The potential impacts of climate variability and change on temperature-related morbidity and mortality in the United States. Environ Health Perspect 109(2):185–189
McMichael A, Woodruff R, Hales S (2006) Climate change and human health: present and future risks. Lancet 367:859–869
Medlock J, Hansford K, Bormane A, Derdakova M, Estrada-Peña A, George J-C, Golovljova I, Jaenson T, Jensen J-K, Jensen P, Kazimirova M, Oteo J, Papa A, Pfister K, Plantard O, Randolph S, Rizzoli A, Santos-Silva M, Sprong H, Vial L, Hendrickx G, Zeller H, Van Bortel W (2013) Driving forces for changes in geographical distribution of Ixodes ricinus ticks in Europe. Parasit Vectors 6(1):1
Meinhardt PL (2006) Recognizing waterborne disease and the health effects of water contamination: a review of the challenges facing the medical community in the United States. J Water Health 4(1):27–34
Milly PC, Wetherald RT, Dunne KA, Delworth TL (2002) Increasing risk of great floods in a changing climate. Nature 415:514–517
Moldoveanu M (2012) Virusologie, bacteriologie si parazitologie pentru asistenti medicali, Ed All, Bucharest, 76-98, 109
National Institute of Statistics, Romania (2014) Romanian statistical yearbook. National Institute of Statistics, Romania, p 278
National Institutes of Health (2016) Climate change and human health. U.S. Department of Health and Human Services. Available online at: https://health2016.globalchange.gov/populations-concern. Accessed in June 2017

Portier CJ, Thigpen Tart K, Carter SR, Dilworth CH, Grambsch AE, Gohlke J, Hess J, Howard SN, Luber G, Lutz JT, Maslak T, Prudent N, Radtke M, Rosenthal JP, Rowles T, Sandifer PA, Scheraga J, Schramm PJ, Strickman D, Trtanj JM, Whung P-Y (2010) A human health perspective on climate change: a report outlining the research needs on the human health effects of climate change. Research Triangle Park, NC: Environmental Health Perspectives/National Institute of Environmental Health Sciences. doi:10.1289/ehp.1002272. Available: www.niehs.nih.gov/climatereport. Accessed in June 2017
Rockel B, Woth K (2007) Extremes of near-surface wind speed over Europe and their future changes as estimated from an ensemble of RCM simulations. Climate Change 81:267–280
Schvoerer E, Massue J-P, Gut JP, Stoll-Keller F (2008) Climate change: impact on viral disease. Open Epidemiol J 1:53–56
Smoyer KE, Rainham DG, Hewko JN (2000) Heat-stress-related mortality in five cities in Southern Ontario: 1980–1996. Int J Biometeorol 44:190–197
Stachel B, Gotz R, Herrmann T, Krüger F, Knoth W, Päpke O, Rauhut U, Reincke H, Schwartz R, Steeg E, Uhlig S (2004) The Elbe flood in August 2002—occurrence of polychlorinated dibenzo-p-dioxins, polychlorinated dibenzofurans (PCDD/F) and dioxin-like PCB in suspended particulate matter (SPM), sediment and fish. Water Sci Technol 50:309–316
Statistics of respiratory tract diseases. Romania. Available online at: http://www.bnm-medical.com/USER-FILES/docs/stat/romania.pdf. Accessed in June 2017
Sterl A, Severijns C, Dijkstra H, Hazeleger W, van Oldenborgh GJ, van den Broeke M, Burgers G, van den Hurk B, van Leeuwen PJ, van Velthoven P (2008) When can we expect extremely high surface temperatures? Geophys Res Lett 35(4):14. https://doi.org/10.1029/2008GL034071.
Tucker MA (2009) Melanoma epidemiology. Hematol Oncol Clin North Am 23(3):383–395
Ulbrich U, Pinto JG, Kupfer H, Leckebusch GC, Spangehl T, Reyers M (2008) Changing northern hemisphere storm tracks in an ensemble of IPCC climate change simulations. J Clim 21:1669–1679
USGCRP (2016) In: Crimmins A, Balbus J, Gamble JL, Beard CB, Bell JE, Dodgen D, Eisen RJ, Fann N, Hawkins MD, Herring SC, Jantarasami L, Mills DM, Saha S, Sarofim MC, Trtanj J, Ziska L (eds) The impacts of climate change on human health in the United States: a scientific assessment. U.S. Global Change Research Program, Washington, DC, 312 pp. https://doi.org/10.7930/J0R49NQX
Wang DZ (2008) Neurotoxins from marine dinoflagellates: a brief review. Mar Drugs 6(2):349–371
Weiss R, McMichael AJ (2004) Social and environmental risk factors in the emergence of infectious diseases. Nat Med 10:S70–S76
WHO (2003a) Climate change and human health – risks and responses. McMichael AJ, Campbell-Lendrum DH, Corvalán CF, Ebi KL, Githeko AK, Scheraga JD, Woodward A (eds), Geneva. Available on line at: http://www.who.int/globalchange/publications/climchange.pdf?ua=1. Accessed in June 2017
WHO (2003b) Climate change and human health – risks and responses. Summary. Available online at: http://www.who.int/globalchange/climate/en/ccSCREEN.pdf. Accessed in June 2017
WHO (2015) The European mental health action plan 2013–2020. WHO Regional Office for Europe, Copenhagen. Available online at: http://www.euro.who.int/__data/assets/pdf_file/0020/280604/WHO-Europe-Mental-Health-Acion-Plan-2013-2020.pdf?ua=1. Accessed in June 2017
World Cancer Report (2014), WHO-IARC Press, Lyon
Wu X, Lu Y, Zhou S, Chen L, Xu B (2016) Impact of climate change on human infectious diseases: empirical evidence and human adaptation. Environ Int 86:14–23
Ziska LH, Caulfield FA (2000) The potential influence of raising atmospheric carbon dioxide (CO_2) on public health: pollen production of common ragweed as a test case. World Resour Rev 12:449–457

Chapter 5
The Energy-Water-Climate Nexus and Its Impact on Queensland's Intensive Farming Sector

Georgina Davis

Abstract Energy and water are inextricably connected in agricultural systems. This chapter discusses the role of climate change on the energy-water nexus and how efforts to increase efficiency in both energy and water end uses can increase Queensland's agricultural sector's resilience. Climate change is continuing to affect water availability and put new stresses on energy systems (particularly in constrained areas), but the degree of future impacts is uncertain, particularly given the changing climate patterns moving towards increasing frequency and the duration of drought conditions, coupled with extreme weather events. Whilst there are a range of technological solutions to improve efficiencies and ultimately productivity, further government and policy support is needed, and this support needs to be coordinated to avoid unintended consequences which have arisen from previous government programmes. This chapter also discusses how efficiency in energy and water end uses can reduce the sector exposure to acute and chronic stressors, including high utility bills which, with climate change, are negatively impacting agricultural productivity. Queensland's intensive agricultural sector stands to gain significantly from an energy-water productivity agenda which acknowledges climate change.

Keywords Energy · Water · Climate · Agricultural systems · Farming sector · Impact

5.1 Background

Queensland is the north-eastern Australian state, covering nearly 1.8 million square kilometres (Queensland Government 2017a), nearly one-quarter of the total land area of the Australian continent, and making it the second largest state. The present

G. Davis (✉)
Griffith School of Engineering, Griffith University, Brisbane, QLD, Australia
e-mail: g.davis@griffith.edu.au

A. Omran, O. Schwarz-Herion (eds.), *The Impact of Climate Change on Our Life*, https://doi.org/10.1007/978-981-10-7748-7_5

Table 5.1 Agricultural statistics for Queensland, Australia

	No. of agricultural businesses	Area of farm (ha)	Total area of crop (ha)
2012–2013	26,647	129,548,236	3,270,474
2013–2014	26,786	139,932,697	2,302,145
2014–2015	25,512	135,917,925	2,407,354
2015–2016	18,153[a]	127,550,998	2,335,703

Source: ABS (2014, 2015, 2016c, 2017b)

[a]*Note*: The scope for the 2015–2016 Agricultural Census is different from previous ABS rural environment and agricultural collections. The threshold for including businesses in the 2015–2016 census was raised from an estimated value of agricultural operations (EVAO) of AU$5000 and over to AU$40,000 and over

population is 4.925 million (Queensland Government 2017b) which is expected to increase to between 8.01 and 11.275 million by 2061 (Queensland Government 2015). Queensland is a temperate and typically sunny state, ranging from tropical to sub-tropical conditions, with average summer temperatures of 29 °C and average winter temperatures around 16 °C. Most of Queensland receives more than half of its annual rainfall during the summer months (December to February) with average rainfall varying from less than 200 mm in the central-west region to over 3200 mm on the far north coast (Australian Government 2017a). The continuing impact of extreme weather events is significantly impacting primary producers across Australia, from crop damage to livestock stress. Climate change is exacerbating the effects of extreme weather events, such as drought, and rural and regional communities are disproportionally impacted (Climate Council 2016). Extreme weather or climate events include heatwaves, bushfires, droughts, tropical cyclones, cold snaps and extreme rainfall (storms, hail, floods), and extreme events are defined as the occurrence of a weather or climate variable above (or below) a threshold value near the upper (or lower) end of the range of observed values (IPCC 2012). These impacts are not uniform across the state.

In 2014–2015, there were 384.6 million hectares of land owned or operated by 123,000 agricultural businesses in Australia (ABS 2016a). Almost half of Australia's total land area is used for agriculture, and, of all the states and territories, Queensland has the highest proportion of agricultural land with 84.3% of the state used for agricultural production (Department of Agriculture and Fisheries 2017). This equates to nearly 136 million hectares of farmland, of which nearly 2.5 million hectares is laid to crop (ABS 2016b). Table 5.1 shows the number of farm businesses, area and total area of crop in Queensland for the past 4 years. Crops contributed AU$27.3b to the total value of agriculture in Australia, with wheat ($6.2b), fruit and nuts ($4.2b) and vegetables ($3.6b) being the largest contributors (ABS 2017a). For 2016–2017, the total value of Queensland's primary industry commodities is forecast to be AU$18.55 million, an increase of 6% from 2015 to 2016 (Department of Agriculture and Fisheries 2017), contributing around 2.5% of the state's economy (Queensland Government Statisticians Office 2016) and accounting for around 24% of Australia's agricultural production value (ABS 2017a), making Queensland a significant agricultural producer particularly when considering the relatively small domestic population.

Despite the changes in reporting practices, the actual numbers of agricultural businesses are declining. Rural Queensland is experiencing population decline in many areas. Technological advances, consolidation and amalgamation of properties to create economies of scale and efficiencies, foreign investment and retirements (where there has been no succession planning), including changing ownership from family- to corporate-owned enterprises, have reduced the number of ratios of farmers to farmland (Climate Change Council 2017). Between 2006 and 2011, there were around 20,000 fewer farmers, a fall of over 11% in just 5 years and a decline of over 40% from 1981 to 2011 (ABS 2012a). Drought and selected declining commodity prices have all contributed to this loss.

Queensland's agricultural industries are comprised of:

- Plant industries, including field crops (sugar cane, cotton, grains and pulses), production horticulture (nuts, fruit and vegetables), lifestyle horticulture (turf, flowers, nursery and landscaping), and forestry
- Animal industries, including livestock and livestock products (including cattle, sheep and pigs, poultry, kangaroos, fish/aquaculture) and livestock products, such as wool, dairy, bees/honey and eggs

Despite Queensland's dominance in primary production, the state 'value-adding sector' (including food manufacturing and processing, distribution and retail) is relatively small compared to other states (Department of Agriculture 2017). Some authors have proposed that this lost opportunity is due to historical factors influenced by Queensland's smaller population as compared to other states, such as Victoria, which has led to a 'commodity culture' as opposed to a 'food processing' or 'value-adding culture' (Woodhead et al. 2017). Whilst this was true of the past, Queensland now has the third largest population. Food processing is increasingly being driven in Queensland by the need to utilise produce that does not meet consumer expectations and, in turn, specification guidelines from the two largest, monopolistic supermarkets; leading to estimated levels of fruit and vegetable waste on Australian farms of between 20 and 40%.

Decisions concerning commodity/crop type are highly dependent on several factors including, but not limited to, climatic forecasts, future water allocations, access to water trading arrangements, electricity tariff (price) options, global commodity prices and fluctuations of various input costs, through to rotational cropping driven by soil health, rising groundwater and other environmental considerations. Farmers are becoming increasingly flexible, adapting to increasing frequency in these fluctuations. For example, sugar cane production in Queensland increased from 30,204,168 tonnes in 2014–2015 to 32,267,975 tonnes in 2015–2016, an increase of 7%, whilst sorghum production decreased by 384,409 tonnes, or 25% to 1,177,451 in 2015–2016 (ABS 2017b). These variations reflect changing commodity prices as well as reduced rainfall for many areas of Queensland. Whilst Queensland's agricultural sector is currently facing a number of major challenges, including climate change highlighted by increasing storm activity and periods of drought, through to competing land use demands, including those associated with the protection of the Great Barrier Reef, the financial value of the sector is growing. It is also seeing diversification across commodities, in particular a move to high-value products and also adoption of technological innovation.

As the global population rises towards the predicted 9.7 billion by 2015 (United Nations 2015), there are increasing demands on, and opportunities for, Queensland's agricultural sector. An increasing global population is driving unprecedented demand for food, fibre and other higher-value products including bioproducts (including biofuels and biochemicals). Queensland is uniquely located in its proximity to Asian markets and is developing new transport and logistic infrastructure to expedite fresh product delivery to growth, often high-end markets. Queensland's agricultural sector is heavily export orientated, contributing approximately AU$9.16b to the state's economy (Department of Agriculture and Fisheries 2016). Whilst it is unrealistic for Australia (let alone Queensland) to be the 'food bowl of Asia', Australian food exports do provide sufficient calorific intake for 61,536,975 people, based on current average daily calorific intake of people in Asia (Australian Farm Institute 2017).

Whilst Australia is an exporter of agricultural commodities (including beef cattle, sugar, cotton, pulses), many of these markets have been established on premium 'Australian-branded' produce for quality assurance and traceability (Woodhead et al. 2015), particularly, high-value-added protein foods, fresh fruits and vegetables through to increasingly niche products, such as guinea fowl.

5.1.1 On-Farm Productivity

In supplying the increasing demand for food in the region, Australia is subject to competition and major institutional impediments. Rather than relying on global markets, most of the world's wealthy industrialised countries have sought to protect their farmers from competition by imposing high import tariffs, import quotas and direct price support mechanisms. By contrast, Australia has withdrawn any producer support in the belief that it would act as a mechanism to drive on-farm innovation and reduce economy-wide costs. The level of agricultural producer support in Australia is currently the second lowest in the OECD area at under 3% of gross farm receipts (ABARES 2014). This approach, at face value, appears to be fundamentally flawed with productivity across Australia's agricultural sector declining. Between 1948–1949 and 2013–2014, the 'total factor productivity' (TFP) grew at an average of 2% per annum, driven by technological advances and innovation (ABRARES 2015). During this period, there has also been a reallocation of resources from inefficient farms to those with higher efficiencies. With research determining that productivity growth was responsible for more than two-thirds of the growth in agricultural output during the post-war era (Mullen 2010). However, over the past 20 years, agricultural TFP has fallen, from 2.6% a year between 1948 and 2000 to 0.9% since the late 1990s (ABARES 2015). Most concerning, TFP declined at −0.2% for the period between 1999–2000 and 2009–2010 (ABARES 2015). The productivity slowdown (and decline) is considered to have been heavily influenced by the Millennium Drought, coupled with declining public investment in agricultural research and development (Sheng et al. 2010).

Productivity growth is essential to maintaining the competitiveness of Australian farmers in the global market as well as to offset negative impacts to on-farm profit factors, such as rising input costs. The sector is managing high production costs (energy, water, labour), excessive regulation, high currency exchange, natural resource pressures associated with climate change and, for some sectors (namely, animal welfare), increasingly multifaceted 'social licence to operate' conditions which has led to a reduction in intensity (e.g. the transition to free-range eggs and poultry). The sector also continues to struggle with fluctuating commodity prices and rising competition from countries with extensive producer support packages. The intensity of this competition is escalating (Keogh 2011); and unlike much of Australia's manufacturing industries (e.g. car manufacturing) which have relocated overseas to benefit from reduced input costs (particularly energy and labour), agricultural land is an 'anchored/in situ' resource. Macquarie Research's estimates that four in ten Australian companies that took their business overseas are expecting significantly better earnings in the future, with lower input costs and higher growth overseas than those solely doing business in Australia.

Agriculture and agricultural productivity sustainably are essential for the long-term welfare and economic security of rural Queensland whilst providing increased food (nutrient) security, both domestically and within the Asia-Pacific area. Queensland's (and Australia's) agricultural sector is showing indications of decreasing capacity and faltering productivity gains, threatening the resilience of some rural businesses. Surveying of rural agricultural businesses by the Climate Council (2016) has shown that many businesses have drawn down on their financial reserves and, in many cases, have taken on increased debt in response to extreme weather events. As Queensland's climate becomes more variable and there are more extreme weather events, adaptation (and cost) to those agricultural enterprises will become increasingly challenging. The ability for many agricultural businesses to put aside/allocate funds for practical on-farm resilience measures, such as technology or infrastructure, is being diminished by unsustainable input costs, most notably the price of grid-supplied electricity. Research (Kiem et al. 2010; Kiem and Austin 2013) into the indebtedness of farming enterprises (food and fibre) has highlighted the financial unsustainability, acknowledging the impacts from rising input costs and the trends in poor profitability over the past decade. This has been coupled with limited number of options other than debt to finance farming activities (McGovern 2014; Moshin 2015).

The Climate Council (2017) refers to climate change as a 'threat multiplier' in terms of its ability to exacerbate existing stresses on rural businesses and communities as well as adding new ones. Queensland farmers now have to manage legacies, such as salinity, which have arisen from historical poor farming management practices (e.g. in the Burdekin region) and new land use challenges such as their co-existence with the new coal seam gas (CSG) industry, a legacy that set of abandoned mines and resulting contamination issues to more recently the failed underground coal gasification plant which has resulted in large-scale soil and groundwater contamination (Department of Environment and Heritage Protection 2017) in a critical agricultural area. The following sections provide a regulatory and policy overview of the various input factors (particularly, water and electricity) influencing agricultural producers in Queensland. Addressing the energy trilemma (secure, affordable and low carbon energy) and ongoing access to water for primary producers has never been more critical at a time of climatic change.

5.2 Climate Variability and Climate Change in Queensland

The term 'nexus' refers to a connection or a connected group or series. The water-energy-food nexus is an increasingly used approach to frame the challenges associated with integrating human development objectives with the responsible management of natural resources (Johnson and Karlberg 2017). This chapter identifies the energy-water-climate change nexus impacting Queensland's agricultural sector, highlighting the interlinkages between input costs (namely, electricity) and the regulatory and policy objectives associated with natural resource planning in an increasingly variable climate.

Queensland benefits from a varied range of climatic zones and microclimates which facilitate a wide diversity of growing conditions and the ability, in some areas, to provide multiple cropping opportunities within a single year. As such, agricultural production is distributed across different climatic zones and also geographically. Much of the most productive arable land is located in the south-east areas or closer to the coast, whilst western parts of the state are dry and drought prone and more widely utilised for grazing purposes.

Northern Queensland is affected by the seasonal migration of the monsoon across the equator. This results in a distinct dry season (May to September) and wet season (October to April), with the latter part of the wet season characterised by monsoonal activity, whereas the south-eastern Queensland region is characterised by a range of climates from sub-tropical in the north through to temperate in the south, with a typically drier winter and wetter summer. This area is also influenced by large-scale atmospheric circulation drivers, particularly the El Niño Southern Oscillation, leading to high variability and the occurrence of droughts and floods (CSIRO 2012). On the eastern Australian coast (Queensland and down into New South Wales), the formation of 'east coast lows' can also have a significant impact on water resources, resulting in intermittent large inflows into water catchments. This area has also been subject to considerable climate variability, including the Millennium Drought and the two wettest years on record for Australia (2010–2011) as a result of two strong La Niña events (CSIRO 2012).

Queensland is the most natural disaster-prone state in Australia (ABS 2012b; Davis 2016). With an increasingly variable climate, impacts to agriculture will manifest, particularly in the frequency and intensity of extreme weather events ranging from longer dryer periods, intense flooding, fewer frosts to warmer temperatures for longer and increased maximums (WWF 2015). Temperatures in Australia were relatively stable from 1910 to 1950. Since then, both minimum and maximum temperatures have shown an increasing trend, with an overall increase from 1910 to 2010 of approximately 0.8 °C (ABS 2012b). Australia has warmed by approximately one degree since 1910. The warming has occurred mostly since 1950, and the frequency of daily temperature extremes has also changed since 1910 (Bureau of Meteorology 2017). The number of weather stations recording very warm night-time temperatures and the frequency with which these occur have increased since the mid-1970s. The rate of very hot daytime temperatures has also been growing since the 1990s (Bureau of Meteorology 2017). These trends are also typical across Queensland (Table 5.2), with various climate records continuously being broken across the state, often with weather patterns within Queensland showing conflicting climatic factors (e.g. drought and flood) during the same periods.

5.2.1 Impacts of Climate Variability and Climate Change to Queensland Agriculture

The Paris Agreement (ratified by 157 countries in November 2016) aims to strengthen the global response to the threat of climate change by ensuring global temperature rise this century which is below 2 °C above pre-industrial levels and to pursue efforts to limit the temperature increase further to 1.5 °C. There has been little research published to date on the assessment of climate change at 1.5 °C and the potential of extreme weather events (Mitchell et al. 2016; Hulme 2016). Climate model simulations (King et al. 2017) have determined that in 'Australia, extreme heat events are projected to become significantly and substantially more common in a 2 °C warmer world than in a 1.5 °C world. Yet the difference in hydrometeorological extremes between a 1.5 °C and 2 °C world is less clear'. For example, extremely wet Decembers such as that seen in 2010 in Queensland are projected to remain rare in a warmer world (King et al. 2017).

There is even less research with regard to the resulting impacts (current and possible future scenarios) for regional Queensland, with most of the research focusing on the highly populated area of south-east Queensland (Hagger et al. 2013; Smith et al. 2013; McDonald 2010). The research and understanding of the existing and future climate change impacts for regional Queensland and, in particular, highly agricultural areas is nascent and requires further research attention as a matter of urgency as higher temperatures and reduced water will impact agricultural production and, in turn, productivity. The Climate Council (2016) has projected a number of climate changes for Australia including, but not limited to:

- Annual average temperature by 2030 is projected to be 0.6–1.3 °C above the 1986–2005 baseline.
- Temperatures by 2090 will be highly variable, based on the success of the Paris Agreement with a low emission scenario temperature range from 0.6 to 1.7 °C and 2.8 to 5.1 °C for a high emission scenario.

Table 5.2 Summary of climate trends for Queensland 2012–2017 (excerpts sourced from Bureau of Meteorology 2017)

2017 (up to 1 August)	Severe tropical cyclone Debbie (23 March to 7 April) was the strongest tropical cyclone (category 4) in the Australian region since cyclone Quang in 2015 and was the most dangerous cyclone to impact Queensland since cyclone Yasi in 2011. Debbie made landfall in north Queensland between Bowen and Airlie Beach
	Ex-tropical cyclone Debbie then caused mass flooding in south-east Queensland
	Queensland had its warmest July on record for both maximum and minimum temperatures across the whole state, with a mean temperature 2.65 °C above the long-term average of 15.94 °C
	Days were record warm, 3.03 °C above the long-term average of 22.95 °C and more than 0.5 °C greater than the previous record set in July 1915
	Nights were warmer than average overall; July minimum temperatures were the tenth-warmest on record, 2.27 °C above the long-term average of 8.94 °C
	Rainfall in the north and northwest for July was above average in northwest of the state
2016	Rainfall in 2016 was well above average in western Queensland and above average in central and southern districts. Conversely, parts of the Gulf country, Cape York peninsula and north tropical coast and areas of south-east Queensland recorded below-average annual rainfall. Both mean maximum and mean minimum temperatures for 2016 were above average for the state
	A dry start to the year in the south-east and far north, particularly from January to April
	Queensland had its 11th driest April on record
	A very wet winter, the second wettest on record for Queensland and the wettest since 1912 with the wettest May to September on record for Queensland
	Flooding occurred from June to September in western, central and southern Queensland
	Queensland's mean temperature was the equal second-warmest on record, at 1.22 °C above average
	Mean maximum temperatures were above to well above average in the north and the east; the mean maximum temperature for Queensland was tenth-warmest on record, at 0.82 °C above average
	Overall, mean minimum temperatures were above to well above average across the state, with the mean minimum temperature for Queensland record warm at 1.61 °C above the long-term average

(continued)

Table 5.2 (continued)

2015	2015 was the third-warmest year on record for Queensland. Rainfall was well below the long-term average in western, central and southern parts of the state
	Exceptional heatwave in March
	Tropical cyclones *Nathan* and *Marcia* make landfall
	Major flooding associated with *Marcia* in the Fitzroy, Burnett and Mary River Catchments
	Below-average rainfall for the state; record low annual rainfall at some locations
	Dry in parts of western, central and southern Queensland, exacerbating long-term rainfall deficient areas
	Severe thunderstorms, hail and flash flooding during January, march and December
	The statewide mean temperature was 1.03 °C above the historical average and 0.22 °C below the record warmest year in 2005
	The statewide mean maximum temperature was the sixth-warmest on record, with an anomaly of 1.04 °C above the long-term average.
	An extended hot spell saw a record number of consecutive days of 40 °C, or above maximum heat records were broken
	The statewide rainfall during 2015 was 487.5 mm, 22 percent below the historical average
2014	Statewide average mean temperature equal-third-highest on record
	Statewide average maximum temperature fifth highest on record
	Heatwaves in January, May, October and November
	Six tropical cyclones: *Dylan*, *Edna*, *Fletcher*, *Gillian*, *Hadi and Ita*
	Flooding in northern, central and western Queensland between January and April
	Below-average annual rain in most areas
	Severe thunderstorms in the south-east during November and December
	Queensland recorded its equal-third-warmest year on record during 2014, with the statewide mean temperature 0.96 °C above the historical average
	Maximum temperature was 1.05 °C above average and the fifth-warmest, whilst minimum temperature was the seventh-warmest and 0.87 °C above the long-term average
	The statewide average rainfall during 2014 was 561.1 mm, 10% below the historical average

(continued)

Table 5.2 (continued)

2013	Mean maximum temperature highest on record with record hot start to the year
	Record rain in the east in January from ex-tropical cyclone *Oswald*
	Four tropical cyclones and severe thunderstorms in November and December
	Dry west of the Great Dividing Range and record low rainfall in the west and southern interior
	Averaged across Queensland, 2013's mean maximum temperature was the highest on record at 1.51 °C above the 1961–1990 average of 29.89 °C. The mean maximum temperature beat the previous 2002 record by a margin of 0.08 °C and was the highest in 104 years
	The mean temperature (average of the maximum and minimum temperatures) was 1.22 °C above the 1961–1990 average of 23.23 °C and the second highest on record
	Following the intense heat at the start of the year, there was a switch to wet weather before the end of January – Category 1 tropical cyclone *Oswald*
	Very heavy rainfall during the period 22–29 January resulted from ex-tropical cyclone *Oswald*, along with extensive flooding for much of the east coast of Queensland. Numerous daily rainfall records were broken during this time; across the whole state, the 489.89 mm was 21% below the long-term average of 623.4 mm
2012	43% of Queensland by area, recorded above-average rainfall, with some areas in the north, west, and south-east (mainly non-coastal) receiving average or, in the case of the south-east and far north-east, below-average rainfall (51 and 6% of the state, by area, respectively)
	Maximum temperatures were a little above average (+0.22 °C), whilst minimum temperatures were below average (−0.36 °C)
	Queensland was the only Australian state to experience a below-average mean temperature during 2012
	The Queensland area-averaged rainfall was 674.4 mm, which is 51.0 mm above the 1961–1990 average of 623.4 mm (lowest since 2008)
	Persistent heavy rain from 23 to 27 January across south-east Queensland triggered extensive flooding which continued in early February with further rain across the southern inland and westward flowing floodwaters progressing across southern Queensland
	A tornado was also reported in the western suburbs of Townsville on 20 March
	Queensland endured a busy fire season, with late October a particularly bad time as severe fire weather saw more than 60 fires burning across the state
	Severe thunderstorms over 17–18 November caused widespread damage in south-east Queensland. High winds, large hail (including up to 10 cm on the Darling Downs), intense lightning activity as well as some flash flooding occurred around Brisbane

Note: There have been three major droughts across Australia within the past century, including the Federation Drought (1895–1903), the World War II Drought (1939–1945) and the Millennium Drought (1996–2010)

- Warming will be greater in inland areas, coupled with a decrease on relative humidity for inland areas.
- Marked southward movement of hot/warm climates corresponding to approximately 900 km movement in climate zones.
- Frequency of extreme droughts to increase for all regions.

- Frequency of heavy rainfall events to increase for all regions, even those where annual rainfall is modelled to decrease, which will increase the incidence of flooding and land/soil erosion.
- More severe/intense cyclones and possible southward extension of cyclone and storm areas.

To fully recognise the additional pressure that climate change poses for rural Queensland, it must be considered in context of the additional challenges (environmental, social, political and economic) already faced by these areas (Brooks and Loevinsohn 2011; Kiem and Austin 2013). Queensland's regional economies, for example, are particularly challenged in terms of short-term boom-bust-type activities. Regional Queensland economies are historically volatile. For example, between 1830 and 1850, wool provided economic gains for regional Queensland; then for another 20 years, gold provided more wealth than wool; and, more recently, 'resource towns' such as Dalby and Chinchilla rapidly expanded on the construction of the CSG industry but contracted quickly once the sector moved into its operation phase resulting in mass sackings and a downturn for house and business prices. Select regional Queensland areas continue to record substantial international and domestic tourist numbers and their associated expenditure (Queensland Government 2017c); however, agriculture remains the consistent employer and economic provider for many rural towns (Rickards 2012). It is an imperative for agricultural systems and businesses to adapt to the impacts of a changing climate which will continue to alter production systems. For example, direct impacts may include variation in plant cycles resulting in different harvesting requirements, increased crop damage from hotter days, reduced crop/produce quality through to impacts on livestock ranging from increased mortality to reduced milk production and changes in quality and quantity of the available 'feed' (Garnaut 2008). Further impacts will influence natural ecosystems which agriculture has come to rely on, including, but not limited to, potentially reduced surface water flows, permanent impacts and slower groundwater recharge conditions through to increased ocean temperatures (impacting open water aquaculture facilities). Already documented examples of direct impacts on Queensland's agricultural sector have included diminished crop/produce cycles (personal information):

- Consecutive heatwaves are impacting strawberry production in south-east Queensland, due to a reduction in flowering as strawberries do not initiate flowers above 29 °C. Increased heat during harvesting also increases the speed of ripening which, in turn, decreases the shelf life of the product and increases the amount of produce rejected by retailers.
- Pineapple farmers on Queensland's Sunshine Coast have experienced reducing product yield and quality due to extreme heat and sunlight and have been applying sunblock (crushed limestone solution) to their fruit to avoid burning. This has increased input costs.
- With Queensland cotton costing approximately AU$1000 a hectare to produce and yield of around 1.5 bales a hectare or less expected and with many of the crops suffering moisture stress, farmers experiencing drier periods are unable to break even.

Indirect impacts such as changing and emerging biosecurity issues (pests and weeds), increased heat fatigue risk to outdoor workers, altered seasonality and increased electricity costs (due to increased pumping requirements) all have a critical impact on agricultural productivity. There may also be interruptions to supply chains, infrastructure and transport structure, particularly during extreme weather events. Some of these risks have been modelled (e.g. Naish et al. 2013), but the effects are so diverse that it is complex to total these models to present an overall or more fully quantified impact. Even more complex to measure are the impacts and losses, including reduced investment confidence and inability or unwillingness to invest in new technologies (Garnaut 2008; Hayman et al. 2012) through to the changes on the social fabric of rural communities, to the impact on mental health of individual farmers and farmworkers (Edwards et al. 2015; Vins et al. 2015).

Queensland (and indeed Australia) has seen much political toing and froing across emission reduction strategies, climate change policies and renewable energy targets and plans, at both state and commonwealth levels over recent years, resulting in lost opportunities and uncertainty. This has undoubtedly cost both business (especially agriculture) and the environment. It is a challenging time for businesses just to keep up with policy changes in this area. However, what is certain for business is the rising liabilities from climate- and carbon-related exposure. This includes increasing electricity costs, rising environmental compliance and compliance obligations and the increasing scrutiny of customers and investors, through to the possible reintroduction of a price on carbon.

5.3 The Regulation of Water for Agriculture in Queensland

Australia's National Water Initiative (NWI) was agreed in 2004 by the Council of Australian Governments (COAG) and provides a national 'blueprint' for water reform (Department of Agriculture and Water Resources 2017a). Under the NWI, state governments and territories across Australia made a commitment to prepare comprehensive water plans to ensure the future sustainability of water (surface and groundwater) reserves and to address any over-allocation or unsustainable 'take' through water entitlements. These water plans, whilst burdensome in terms of regulation, have provided certainty of water resources to farmers and facilitated improved crop and on-farm planning for many agricultural businesses, allowing for greater investment in infrastructure and on-farm improvements.

Policy guidelines for the NWI were established to assist jurisdictions and develop and implement NWI-consistent water planning and management arrangements. In 2017, to support the NWI policy guidelines, new guidance to jurisdictions on how to consider and incorporate possible impacts from climate change and extreme events in water planning and management was published (Department of Agriculture and Water Resources 2017b). The guidance acknowledges that climate change and extreme events are significant challenges for governments at all levels, as it causes changes in water availability and reliability in some regions and in the frequency and severity of droughts. This ultimately impacts on the environment and agriculture and may even result in direct damage to water infrastructure by more extreme weather events such as floods, cyclones and bushfires. The NWI identifies water access entitlement holders (e.g. an agricultural landowner with a water entitlement) as responsible for bearing the risks of any reduction in water allocation, including the reliability of water allocation, resulting from seasonal or long-term changes in climate and also any periodic natural events such as drought (see NWI 2004: clause 48). This approach aligns with a key 'driver for good adaptation' where climate risks are well understood and clearly allocated to those best placed to manage them (DIICCSRTE 2013).

Queensland has enshrined its NWI obligations into the *Water Act 2000* (Qld) and has developed 24 water planning areas (Department of Natural Resources and Mines 2017). Water plans for these areas provide a legal 'head of power' as they are subordinate legislation and apply to surface water and, in some areas, groundwater. The adoption of approaches in each plan varies depending on local circumstances. These include expected climate change impacts, the level of development of the water resource and the type of water supply system (e.g. groundwater or surface water, connected or nonconnected system).

Queensland's climate variability means that it is difficult to identify how water resources may be affected over the timeframe and scale of a water plan, and, as such, water plans are able to respond to a range of intraannual and interannual variations. The Queensland Government uses a range of hydrological and climate modelling that incorporates historical climate variability and future projections (where relevant) to determine sustainable water use and to inform water entitlements and has incorporated adaptive approaches to manage water resources. The natural expiry (every 10 years) of the plans then permits for any readjustments to water entitlements from climate change to be made whilst providing a period of certainty to the agricultural water user. The science collected over the life of the water plan is essential information for evaluating the effectiveness of implementation and whether a plan is achieving its outcomes. This legislation also provides a framework for a trade-off between environmental flows and consumptive water. Changes in water resource availability may require reconsideration of water made available for human uses and that set aside for the environment. For example, if water availability declines due to climate change, the relative allocation of water for consumptive versus environmental use will reflect a complex mix of allocation rules which are contained within Queensland's legislation and also the water plans themselves. In Queensland, there are requirements so that water for human needs and minimum environmental flow conditions is met before allocation of water for general use, including for agriculture.

5.3.1 Agricultural Water Use in Queensland

Queensland has a range of water systems which are utilised by agricultural water users and which may be regulated or controlled through different mechanisms. The type of water supply system, for example, surface water (including irrigation schemes) or groundwater (or both), and whether the system is large and connected or small and nonconnected, also impacts the range of approaches used to respond to climate risk. Large surface water systems with a connected distribution system are generally well suited to water trading, which can be an effective approach for managing climate risk. Whilst small self-supply, nonconnected surface water systems are often regulated through allocations which may be 'high', 'medium' or 'low reliability'. Groundwater systems, whether large or small, require additional considerations for groundwater-dependent ecosystems and ensure the groundwater resources receive adequate recharge. In regions where a drying climate is expected, planning for a reduction in water extraction may be required. The total volume of water used by Australia's agricultural sector in 2015–2016 decreased by 3% to 9,157,291 megalitres over the previous year, but the total area watered across Australia increased by 4% to 2,147,915 hectares. Table 5.3 shows the water use by Australian agriculture by state or territory and the range of sources of the agricultural water. Queensland has the second largest number of irrigated agricultural businesses in Australia and is the largest user of water in agriculture. And whilst Queensland is the largest user of recycled/recaptured water, it is also the largest user of groundwater resources. It is noted that the amount of energy and, in turn, the financial cost of utilising (pumping) groundwater (electricity, diesel) are higher than utilising surface waters.

Table 5.4 shows a summary of agricultural water use in Queensland across a 7-year period, showing a discrepant irrigation profile. However, the amount of water used from irrigation channels and groundwater sources has, for the most part, been steadily increasing. It is important to note that the Queensland climate over this period has been highly variable, with many areas experiencing periods of flood or drought or, in some cases, both.

Climate change is altering rainfall patterns across Queensland. In the areas which are becoming dryer and, in many cases, also warmer, sources of agricultural water from run-off, surface water and groundwater recharge are all being negatively impacted. There have also been significant losses in man-made water systems from on-farm storages to dams. However, the water storage levels are quite discrepant across Queensland. In south-east Queensland, there are the 'major twelve' dams which contribute to the water supply system. These dams have had water storage levels in excess of 70% for the past 12 months. However, just outside the capital city of Brisbane, in the major horticultural region of the Lockyer Valley, the three dams in this area have had significantly reduced levels (Atkinson 6%, Bill Gunn 9%, and Lake Clarendon 8%) (SEQWater 2017). A significant number of water storages (dams and weirs) in regional Queensland are at low levels (50% and under) with many of those dams heavily utilised for agricultural purposes, such as Leslie Dam and Bjelke-Petersen Dam showing the lowest storages (SunWater 2017a). Future projections for the next 12 months show many of these storages declining further (SunWater 2017b).

Table 5.3 Water use on Australian farms for the year ending 30 June 2016

	Aust.	NSW	Vic.	Qld	SA	WA	Tas.	NT	ACT	MDB[a]
Agricultural water use										
Agricultural businesses ('000)	85.7	26.1	20.8	18.2	9.5	8.4	2.3	0.4	0.0	35.5
Agricultural businesses irrigating ('000)	22.7	5.3	6.0	5.4	3.1	1.5	1.2	0.2	0.0	9.2
Total water use ('000 ML)[b]	9157.3	2805.3	2095.0	2646.1	858.8	372.6	332.1	47.0	0.3	5209.9
Water applied for irrigation ('000 ML)[c]	8381.4	2610.9	1946.1	2433.5	777.8	287.5	308.7	16.9	0.1	4938.4
Water applied for other agricultural purposes ('000 ML)[d]	775.9	194.5	148.8	212.6	81.0	85.1	23.5	30.1	0.3	271.6
Change in total water use from 2014–2015 (%)	−2.5	−15.5	−10.2	11.3	15.8	16.7	36.6	−21.9	32.5	−12.6
Sources of agricultural water										
Irrigation channels or pipelines ('000 ML)	3096.2	965.7	1067.7	789.2	117.9	101.2	53.2	1.3	0.0	2087.6
On-farm dams or tanks ('000 ML)	980.1	228.6	103.1	410.5	20.1	87.6	126.4	3.6	0.2	438.5
Rivers, creeks or lakes ('000 ML)	2412.5	885.2	540.0	610.1	231.4	14.1	124.0	7.5	0.1	1656.4
Groundwater ('000 ML)	2357.2	675.5	292.3	761.7	432.3	141.6	19.3	34.4	0.0	926.2
Recycled/reused from off-farm ('000 ML)	161.0	23.3	49.9	57.3	11.3	15.6	3.6	0.0	0.0	56.2
Town or country reticulated mains supply ('000 ML)	126.5	21.3	37.5	8.8	43.6	11.8	3.4	0.1	0.0	35.6
Other water sources ('000 ML)	23.8	5.7	4.5	8.4	2.2	0.7	2.2	0.0	0.0	9.5

Source: ABS (2017c)

Note: The Queensland Murray-Darling Basin (QMDB) comprises the border rivers, Moonie, Condamine and Balonne, Nebine, Warrego and Paroo catchments, all of which have water plans. There is also a Murray-Darling Basin Plan (Basin Plan) which provides a framework to manage the water resources of the basin and sets out reductions in water diversions for basin catchments. The Basin Plan (Cth) requires all basin states including Queensland to undertake monitoring and reporting as well as prepare commonwealth-accredited water resource plans

[a]Murray-Darling Basin (MDB)

[b]Includes water applied for irrigation and other agricultural purposes

[c]Includes water applied to pastures and crops

[d]Includes livestock drinking water, dairy or piggery cleaning, etc

Table 5.4 Summary of agricultural water use statistics for Queensland, Australia

	2015–2016	2014–2015	2013–2014	2012–2013	2011–2012	2010–2011	2009–2010
Area of holding (ha)	127,55	135,91	139,93	129,54	137,23	139,83	129,66
	0,908	7925	2697	8236	9082	4696	7586
Number of agricultural businesses irrigating (no.)	5416	7622	7461	6685	7572	8023	9402
Water taken from irrigation channels or irrigation pipelines – total volume used (ML)	789,231	774,520	774,054	661,279	523,184	373,595	825,248
Groundwater (e.g. bores, springs, wells) – total volume used (ML)	761,703	631,089	575,703	467,064	431,422	424,561	537,691
Total area watered (ha)	529,637	514,467	551,585	514,986	492,382	474,844	502,600
Total application rate (ML/ha)	4.6	4.3	5	4.6	3.8	3.6	3.6
Total volume applied/used (including other agricultural water) (ML)	2,646,093	2,467,277	2,957,845	2,623,228	2,108,250	1,693,994	1,823,870

Source: (ABS 2011, 2012c, 2013a, 2014, 2015, 2016c, 2017b)

Farmers are modifying their practices to adjust to water availability and climatic conditions as above-average temperatures, and dry conditions in Queensland's drought period, along with increasing high prices for water and the electricity to pump that water, are critical factors not only in water use and crop selection but in the 'decision to plant'. This may include 'selling off' water allocations to recoup costs rather than cropping in a bad year, watering larger areas less or changing crops planted. The increasing climate variability across Queensland has highlighted the need to further address water availability, water transfer (trading) and water storage to ensure economic growth and expansion of agricultural regions and new opportunities. Queensland's irrigators experience a range of challenges from reduced water availability, be it from high water prices, climate change through to increased competition for existing available consumptive water (Khan et al. 2006). In Queensland, this competition is from other agricultural users, power stations through to the extractive mineral and resource industries. Another challenge is the rising cost of electricity which was identified as an issue as early as 2008 (Zilberman et al. 2008).

5.4 Electricity Regulation and Pricing in Queensland

Electricity prices in Australia are higher than overseas jurisdictions (CME 2012), disadvantaging commodity exports on the global market and leaving agricultural producers heavily trade-exposed. As Queensland's electricity costs rise, the viability of irrigated agriculture businesses is being eroded. Queensland's electricity prices doubled between 2007–2008 and 2013–2014, predominantly driven by increases in network charges which increased sixfold from 2004–2005 to 2014–2015, accounting for over 95% of the total electricity price increases during the period. Network charges now account for over half of Queensland's retail electricity prices, whereas in 2004–2005, they only accounted for around 20%. By contrast, generation and retail costs remained relatively stable over that period, until 2015. Queensland's average wholesale prices have doubled over the past 2 years (2015–2017) from $52/MWh to $95/MWh, resulting in Queensland having the second highest average wholesale prices in the Australian National Electricity Market. Even the Queensland Government did not forecast these market changes. In 2016, Queensland Productivity Commission review failed to predict these increases, assuming only a 2.1% increase to wholesale prices, rather than the actual 40%. Specifically, irrigation electricity tariffs in Queensland have risen over 136% over the past decade (Table 5.5), and post-2020 this rise will be unsustainable with the withdrawal of these specific, 'non-cost reflective' (and thus transitional) irrigation tariffs. There are about 42,000 electricity connections for businesses in regional Queensland; almost a third of regional business connections are on eight different tariffs classified as transitional or obsolete. Almost half of connections are for agricultural purposes. The calculated bill impacts post-2020 are (Queensland Productivity Commission 2016*)*:

- Tariff 62 – over 50% of the 8800 small customers are going to experience a significant bill increase, and 93.8% of the 290 large customers would be worse off.
- Tariff 65 – over 40% of the 4900 small customers are going to be considerably worse off, and 98.4% of the 100 large customers.
- Tariff 66 – almost 30% of the 2900 small customers are going to be considerably worse off, and 100% of the 100 large customers.

According to the Australian Energy Regulator (2017), for the first three quarters of 2016–2017, there have been 556 small business disconnections in regional

Table 5.5 Electricity prices and irrigation tariffs (Queensland Competition Authority 2017): prices in Australian $

Units c/kwh unless specified otherwise	2017	2016	2015 and 2014	2013	2012	2011	2010	2009	2008	2007
T62										
First 1000 KWh/ month	46.516	44.259	39.411	35.828	32.571	29.61	27.78	24.52	21.93	20.11
Remaining KWh	39.336	37.427	33.228	30.298	27.544	25.04	23.49	20.73	18.54	17.00
All usage (other times)	16.448	15.650	13.939	12.669	11.517	10.47	9.82	8.67	7.75	7.11
Service fee (cent/ day)[a]	78.451	74.644	66.468	60.425	54.932	$15.20/month	$14/26/m	$12.59/m	$11.26/m	$10.32/m
T63										
First 1000KWh/ month					63.036	52.53	49.28	43.50	38.90	35.67
Next 9900KWh/ month					38.580	32.15	30.16	26.62	23.81	21.83
Remaining KWh					30.216	25.18	23.62	20.85	18.65	17.10
All consumption					13.296	11.08	10.39	9.17	8.20	7.52
T64										
All usage					30.792	25.66	24.07	21.25	19.00	17.42
Usage other times					16.908	14.09	13.22	11.67	10.44	9.57
Min. payment per day (cent, c)					53.618	$13.60/month	$12.76/m	$11.26/m	$10.07/m	$9.23/m

T65										
All usage	36.894	35.305	31.438	28.580	25.982	23.62	22.16	19.56	17.49	16.04
Usage other times	20.321	19.446	17.316	15.742	14.311	13.01	12.20	10.77	9.63	8.83
Service fee (cent, c)[a]	78.003	76.644	66.468	60.425	54.932	$15.20/month	$14.26/m	$12.59/m	$11.26/m	$10.32/m
T66										
First 7.5 kw ($ per kw)	37.503	$35.888	31.957	29.082	26.411	24.01	22.56	19.88	17.78	16.30
Remaining kw ($per kw)	112.759	$107.903	96.085	87.350	79.4.9	72.19	67.72	59.78	53.46	49.02
Energy charge										
All usage	19.338	18.505	16.478	14.980	13.618	12.38	11.61	10.25	9.17	8.41
Service fee (cent, c)[a]	171.915	164.512	146.493	133.175	121.068	$33.50/month	$33.50/m	$22.74/m	$24.81/m	$22.75/m

Note:
Tariff 62, farm – time of use (for use between 7 am and 9 pm)
Tariff 63, farm – time of use (phased out 2012–2013)
Tariff 64, irrigation – time of use (agreed 12-hour period, phased out 2012–2013)
Tariff 65, irrigation – time of use (agreed fixed 12-hour period)
Tariff 66, irrigation
[a]Service fee is based on per metre point per day unless otherwise stated
2014–2015 shows the political decision by the LNP government to freeze prices

Queensland alone, which is on track to more than double the 384 disconnections recorded in 2015–2016. Some of these disconnections have been agricultural businesses who could not pay their electricity bill. Irrigators' decisions to utilise irrigation and therefore use electricity are driven primarily by crop water requirements and regulation governing water access (e.g. water licencing conditions which may be based on specific times of the day through to flood levels in a riverine system). Due to these constraints, irrigators often have limited flexibility in their electricity use and cannot respond to different electricity price signals (such as peak versus shoulder electricity rates). In response to these price increases, farming businesses, including irrigators, have been installing energy efficiency measures and renewable energy and in many cases simply reducing demand. However, energy efficiency gains have been diminished by the increasing costs, whilst simply reducing demand has also come at a cost either through reduced productivity or through to farmers simply choosing not to plant a new crop.

Many farmers are now weighing up options to 'switch off' efficient irrigation technologies or leave the grid, taking opportunities in advancing technologies and their reducing costs. However, due to irrigation demands, through to the need for continuous power to refrigerate produce, many are looking at hybrids of renewables and new diesel generation. Whilst diesel presents an attractive option currently given its relatively low-cost and high-reliability, there is future uncertainty on how diesel may be impacted by Australia's obligation to manage carbon. This also leaves a legacy for those customers who are unable to leave the grid and may have to pay increasing costs into the future, thus compounding negative outcomes.

The energy sector has and continues to undergo technological disruption, both in physical technologies for the generation, storage and use of power and in 'soft' technologies to monitor, manage and securely trade power. New technologies and business models offer cheap, reliable energy and are encouraging regional customers to consider going off-grid. The challenge for the Australian Energy Market (and its associated regulation) is to adapt quickly to accommodate these new technologies to retain and attract customers. Indeed, without better incentives, the CSIRO and Energy Networks Australia (ENA) predict 10% of customers are likely to leave the grid by 2050 (CSIRO and ENA 2017). Grid utilisation will continue to decline leading to further regulatory intervention, higher costs and the electricity grid's 'death spiral' (Simshauser 2017). The current electricity grid is competing against a range of technologies that enable consumers to cost-effectively opt out of grid-supplied power. The market needs to move to a customer-driven model allowing consumers to interact directly with the network rather than the current supplier-customer model.

Technology already exists to allow electricity to be securely traded through a genuine net-metered or peer-to-peer trading model. Trading power between nearby sites whilst paying only for 'local network use' may increase grid utilisation and stability, as customers install optimum generation and storage across their sites rather than overcapitalising in plant at individual sites to meet off-grid requirements. It can also offer new network revenue opportunities whilst encouraging customers to stay grid connected. This technology is beneficial for farms with multiple network connections where renewables can be connected to the main account and generation credited against consumption at multiple irrigation pump connections, for example.

More than any other sector of the economy, irrigated agricultural productivity in Queensland is highly dependent on seasonal variations in rainfall and access to a reliable water supply which, in most cases, can only be secured through a sustainable energy (electricity) supply. Changes to weather patterns are influencing both the intensity and duration of rainfall and thus redefining the suitability of many areas for farming and resulting in many irrigators having higher-than-average load factors compared to other energy consumers. For some farmers, changes to rainfall patterns and water shortages will inevitably mean surrendering their farms as production falls, and the level of farm indebtedness becomes unsustainable. The government therefore has a role to assist agribusinesses to manage this risk and ensure future food security. Significant investments in infrastructure and technology and growing innovation across the sector will provide some opportunity, but as agriculture is and will always remain a high-risk industry, the sector often fails to attract the required investment capital. Historically, farmers have responded to their eroding terms of trade by increasing productivity – in many cases this requires access to water which, in turn, can only be achieved and guaranteed by the corresponding access to affordable, reliable energy which will assist Queensland's agricultural sector to compete on the global market.

5.5 Case Study: Queensland's Murray-Darling Basin and the Direct Relationship Between Climate, Energy and Water

The Murray-Darling Basin (MDB) is the most significant river system, covering 14% of Australia's land area across Queensland, New South Wales, Victoria, South Australia and the Australian Capital Territory. The geography of the Basin takes in 23 rivers and their catchments. Almost four (4) million people inside and outside the Basin need its water for the survival of their families, communities and industries. Around 70% of Australia's irrigation occurs within the Basin, accounting for 40% of Australia's agricultural produce (MDBA 2014). The MDB is a diverse and dynamic system. Its characteristics are constantly changing in response to the influences of people, the climate and the way water is used for production, by communities (including to meet the cultural values of the traditional people) and the environment. Water availability is a rising issue for the Basin, and the Millennium Drought (1996–2010) took its toll on the system, in particular the volume of available water for both consumptive use and for environmental flows. In 2012, the Murray-Darling Basin Plan (the Plan), which is a legislative measure (Australian Government 2017b), was released. Its aim was to ensure that water is shared between all users, including the environment, in a sustainable way. The Plan achieves this by managing the Basin as a single system, which is both political and practically complex given the Basin's presence across multiple jurisdictions. There has also been the establishment of the Murray-Darling Basin Authority (MDBA), tasked with promoting the sustainability of the Basin and addressing areas of oversupply (MDBA 2014).

The Plan was originally designed to return 2750 gigalitres per annum of consumptive water (about one-fifth of existing entitlements) to environmental flows within the Basin to protect and restore valuable ecosystems. It also outlines water sharing mechanisms to cope with current and future climate change (Reisinger et al. 2014). Climate change is predicted to decrease water availability in the Basin, posing a risk to the agricultural businesses which rely on the water and have built businesses and infrastructure on the availability of that water (Reisinger et al. 2014), not to mention the communities and homes which rely on the Basin for their drinking water. Indeed, agricultural production within the Basin has been reduced and will be further reduced if the climate change predictions (which predict a substantial 'drying') are realised, despite comprehensive adaptation (Garnaut 2008, Qureshi et al. 2013).

Strategic buy-backs of large volumes of water through the Plan have been the leading tool to address over-allocation and the reductions in water flow due to climatic changes. For Queensland, the impacts of these buy-backs have been mixed, with environmental water purchases increasing water prices to unsustainably high levels during the peak buying period, although this, in part, was also attributable to a low level of confidence in the agricultural sector due to widespread drought conditions through the buy-back period. Rural communities which have heavily relied on the MDB for (irrigation) water are seeing that (irrigation) water diverted to maintain environmental flows to restore ecosystem health (Gell and Reid 2014). Agriculture and the communities in the northern Basin (of Queensland) have a high dependence on irrigation. These communities rely on the existing water allocations and have been already impacted by operational, technological and market changes which have resulted in demographic and social pressures. These communities are also experiencing other challenges, such as exponential increases in the cost of electricity through to the loss of local essential local welfare services. As such, community vulnerability to further water buy-backs, either willing or mandatory, is high. Peak advocacy groups (Queensland Farmers Federation 2017) have called for the need for communities to be appropriately informed and equipped with the right tools and assistance to be adaptive and resilient to these changes. They have also called for the adoption of non-flow, complimentary measures into the assessment approach to meet the 'Sustainable Development Level' and improve overall environmental factors including the control of pest species such as carp which reduces water quality, through to cold water pollution mitigation through the installation of thermal curtains on major headwater storages.

More work must be done by the Queensland and Australian Governments to understand the environmental improvements and water equivalence of these measures. Particularly the future of the buy-back of irrigation entitlements from 'willing sellers' is in doubt with many of the willing sellers now removed from the market. In these cases, prime agricultural land has been taken out of production. Farm businesses are also now buying water to simply 'give back' later so that they can assure their future existence. These perverse, negative outcomes are unsustainable. The MDBA has also yet to address how it will make the water recovery target should there be no more willing sellers. The agricultural sector must have certainty that there will be no mandatory buy-backs or mandatory recovery into the future – otherwise there will be no investment confidence or productivity improvements.

5.5.1 Energy-Water Nexus Issues in the MDB

The commonwealth government is permitted to buy-back a maximum of 118 gigalitres in Queensland. Under this maximum level or cap, the Australian Government also prioritised water infrastructure programmes to assist in the delivery of the water recovery targets. This was done through the 'Healthy Headwaters' programme in Queensland, which provided farm subsidies to upgrade irrigation infrastructure, with irrigators contributing at least 10% of the cost as well as at least 50% of the water savings (by permanent transfer of water allocation) to the Australian Government for environmental use. To date, more than 80 Queensland projects are being progressed, representing a water saving of over 46 gigalitres and government funding of more than AU$110 million to irrigators.

An opportunity for improving water use efficiency of irrigation is to replace gravity-fed irrigation systems, such as furrow or border strip, with more efficient pressurised systems (Jackson et al. 2010). The Healthy Headwaters programme was established to increase water efficiency of on-farm irrigation through the substitution of gravity-fed systems to pressurised irrigation methods. It is now recognised that many of the high-pressure, water-efficient irrigation equipment uses much more electricity than the former, often flood irrigation methods. However, the associated impact from these 80 projects on energy use (electricity) on farm is not considered as it falls outside the scope of the programme. Optimising one aspect of the irrigation process, without due consideration of all other inputs and productivity factors, has resulted in unintended resource and environmental outcomes. Utilising less water on-farm may also not equate to water savings across the Basin (Perry et al. 2009; Molden et al. 2010).

Irrigation is the largest consumer of energy (electricity and diesel) on Queensland's farms (Queensland Farmers' Federation 2017; Davis and Chamberlin 2016). The energy required for pumping (water) depends on crop requirements (which rise during periods of dry, hot weather), pump type, size and efficiency and, for groundwater, the total dynamic head (height that a fluid is to be pumped, taking account friction losses in the pipe) and distance to in-field application. Studies which have investigated the impacts of the transition from flood irrigation to pressurised systems within the MDB (Jackson et al. 2010; Jackson et al. 2011) have determined that whilst there was a reduction of between 10 and 66% in the amount of water applied on-farm, energy consumption increased by up to 163%.

The highest energy increases were seen in surface water systems, whereas energy efficiencies could be achieved from groundwater application noting that "where groundwater is used for irrigation, converting to pressurised micro-irrigation systems can decrease energy consumption if the conversion means that the operating pressures and pumping volumes are reduced" (Jackson et al. 2010). Energy savings are possible when converting from gravity-fed systems to pressurised irrigation methods in groundwater areas, as there is a reduction in the volume of water being pumped (Jackson et al. 2010). However, Jackson et al. (2010) also note that excess water applied from gravity-fed irrigation methods, which was not consumed by the crop/plant, is subject to reuse and, subject to geological conditions, will likely drain to shallow aquifers or back to surface waters. The recharge effects of gravity irrigation schemes are not taken into account, and there has not been any qualification of the consumed and nonconsumed fraction in the MDB.

Flood irrigation techniques are also an adopted strategy for managing groundwaters, particularly in areas of rising saline groundwaters. For example, in the late 1970s, Queensland's Burdekin region saw the finalisation of a surface water irrigation channel scheme. With additional water supplies more readily available, the Lower Burdekin saw a dramatic increase in surface water irrigation in the late 1980s which led to increased groundwater recharge and rising groundwater tables with some parts of the Lower Burdekin experiencing a rise in groundwater levels of up to 10 m over the last 20 years. This has resulted in groundwater levels at less than 3 m below the ground surface across approximately 15% of the irrigated area within the project area. The situation is more acute in some areas where the groundwater table has been measured at only 0.5 m below the surface (Queensland Government 2017d). High water tables can result in water logging of the soil profile and can also mobilise salts from the underlying bedrock which can increase salinity levels. Both of these factors can reduce the productivity of agricultural land and limit opportunities for future development. High groundwater levels can also lead to higher rates of property and catchment run-off flowing into downstream receiving environments.

5.6 Discussion

Framing an energy-water-climate change 'nexus' for Queensland's agricultural sector allows for the identification of critical interlinkages between natural resource management and factors such as agricultural productivity and provides an opportunity to identify the implications from uncoordinated decision-making or government programmes. Domestic policy settings are critical determinants of agricultural productivity as they shape farmers' incentives and capacity to innovate and improve productivity. The imperative of a strong, sustainable and resilient agricultural sector is essential to provide social and economic value to Queensland's rural areas and provide food security. As such, there must be state–/economy-wide agricultural policy settings which create conditions conducive to innovation to ensure an efficient and effective agricultural sector.

Queensland's agricultural sector is vulnerable to global climate change impacts due to its geographical location, geological attributes through to its reliance on climate-sensitive receptors. Queensland already has a regulatory framework for sustainable water use to manage both water quality and water quantity, which are vital to the future agricultural production and maintenance of environmental flows. Furthering water productivity and improving energy use efficiency are critical pathways to reducing the environmental footprints of agricultural production (Khan et al. 2009) and reversing productivity reductions. The adoption of best practice management approaches for land and other resources (water, energy) is an integral component of sustainable agriculture. However, the rising cost of electricity in Queensland continues to outpace the energy efficiency savings and productivity improvements that agriculture can achieve. This is impacting the ability for irrigators (and nonirrigated agriculture) to compete on a level playing field with overseas producers.

The continuous increases in electricity prices coupled with reduced reliability is resulting in more agricultural businesses increasing their use of diesel. According to 2011 data (ABS 2013b), diesel accounted for 81% of agricultural energy use, equating to 76% of total sector annual energy cost. However, more recent data (National Centre for Engineering in Agriculture 2015) indicates diesel is now 87% of the energy cost (noting the reduction in diesel prices, between 2011 and 2015), and agricultural businesses are moving 'off-grid' in regional areas. Whilst electricity prices remain high and the reliability of electricity delivered via the grid diminishes, the value proposition offered by the traditional network is decreasing. This is not only impacting the long-term viability of the grid but is increasing the associated carbon emissions with on-farm energy usage – with carbon emissions contributing to climate change. The Queensland Government must address (through evidence-based strategies) the future of the electricity distribution networks in regional Queensland and provide future policy certainty to users.

Australian decision-makers have eluded that innovation "is the main driver of farm-level productivity growth, as farmers reduce costs by adopting more efficient technologies and management practices" (ABARES 2014, p3). However, this is overly simplistic. Adoption in new, innovative technologies has largely been driven by the lack of sustainability of the 'business as usual' approach. Farmers who have adopted/implemented innovation (particularly in energy efficiency) have found that any potential savings that were identified in the project planning stage have been subsumed in the unsustainable increasing cost of electricity.

With some farmers increasingly choosing not to plant a crop thus reducing their exposure to these prices, particularly where they are able to sell water allocations or, in some cases, their land to often competing land-uses (such as domestic housing or commercial premises, or increasingly to renewable energy applications, such as large-scale photovoltaic facilities). The maturing resource industries, particularly the coal seam gas industry and, more recently, the renewable energy industry in Queensland, are offering some farmers a guaranteed de-risking of their businesses, and, in some cases, farmers are receiving sufficient financial incomes from not farming. For example, across Australia, approximately AU$20.6 million is paid annually in lease payments to farmers and landholders hosting wind turbines (Climate Council 2016), whilst in Queensland, AU$400 m is being paid to 100 landowners for hosting gas wells on their property (TSBE 2017).

All of this, at a time where Queensland farmers are competing with countries with similar climates and soils and with far-reduced input costs, lower levels of regulation and significant tax-concessions for both foreign and domestic investment. The 'playing field' is far from fair/level, and innovation alone cannot fix the systematic and policy failures impacting Queensland's agricultural sector. As climate variability and extreme weather events increase, more Queensland agricultural businesses will need to implement more transformational adaptive or substantial adaptation options, which often may carry more risk and expense, particularly for individual farmers. With some input costs reaching unsustainable levels, the ability for many farmers to financially invest in 'future-proofing' their agricultural businesses and lifestyle is being irreversibly threatened. Even slower, incremental changes and those often referred to as 'low hanging fruit' are out of reach to some farmers based on disposable incomes and the 'opportunity cost' of existing funds.

5.7 Conclusion

There are already signs of productivity fatigue and environmental stress throughout Queensland's agricultural sector. There are both continuing (growing populations, export markets) and new pressures (biofuels and bioproducts) on Queensland's agriculture to produce more whilst safeguarding the natural ecosystems and resource base (particularly water). Much of the additional production must come from the intensification of land and water consumption currently under productive use. This is becoming more challenging against an increasingly changeable climate. A focus on the provision of evidence and modelling data in making innovative and adaptive decisions is critical for agricultural businesses. These models have important implications in policy planning and development towards minimising the impact of climate change on farming practices through to biosecurity risks. Government departments and funding should facilitate knowledge and information access, so that agricultural businesses and their advisors can frame climate change adaptation actions and policies which focus on driving productivity, seizing opportunities and at the same time managing risks, particularly those associated with declining natural resources, rising input costs and increasing regulation.

Energy and water efficiency programmes should be integrated – targeting efficiencies in water, energy *and* carbon to demonstrate water, energy and climate resilient design (e.g. future operational costs and a climate change adaptation). This will require consistent and holistic cross-government agency and multidisciplinary policy to address the competing demands on natural resources (including water and energy) whilst maintaining productive farming systems. Rural communities across Queensland are naturally resilient. Many of landholders and rural communities have substantial experience of extreme weather events and climate variations, but these challenges are escalating, and more innovative and coordinated adaptation is required than ever before.

References

ABARES (2014) Australian Bureau of Agricultural and Resource Economics and Sciences. Australian agricultural productivity growth: past reforms and future opportunities report authors Emily Gray, Max Oss-Emer, Yu Sheng research report 142, February 2014

ABARES (2015) Australian Bureau of Agricultural and Resource Economics and Sciences. A manual for measuring total factor productivity in Australian Agriculture Report authors Yu Sheng and Tom Jackson technical report 152, October 2015
ABS (2011) Australian Bureau of Statistics. Catalogue no 46180 water use on Australian farms 2009–10 released 11 May 2011
ABS (2012a) Australian Bureau of Statistics. Australian farming and farmers 41020 – Australian social trends December 2012
ABS (2012b) Australian Bureau of Statistics. 1301.0 – year book Australia, 2012
ABS (2012c) Australian Bureau of Statistics. Catalogue no 46180 water use on Australian farms 2010–11 released 29 June 2012
ABS (2013a) Australian Bureau of Statistics. Catalogue No. 4618.0. Water use on Australian farms 2015–16. Released 31 May 2013
ABS (2013b) Australian Bureau of Statistics. Energy accounts 2011–2012. Catalogue No 4606
ABS (2014) Australian Bureau of Statistics. 71210DO003_201213 agricultural commodities, Australia, 2012–13. See Table 4: agricultural commodities, State and NRM Region–Queensland–2012–13
ABS (2015) Australian Bureau of Statistics. 71210DO004_201314 agricultural commodities, Australia, 2013–14. See Table 4: agricultural commodities, State and SA4 Region –Queensland–2013–14
ABS (2016a) Australian Bureau of Statistics. Land management and farming in Australia, 2014–15. Report number 4627.0. Released 25/05/2016
ABS (2016b) Australian Bureau of Statistics. Agricultural commodities, Australia, 2014–15. Report number 7121.0. Released 23/03/2016
ABS (2016c) Australian Bureau of Statistics. 71210DO003_201415 agricultural commodities, Australia- 2014–15. See Table 4: agricultural commodities, State and SA4 Region – Queensland–2014–15
ABS (2017a) Australian Bureau of Statistics. Catalogue No. 7503.0. Value of agricultural commodities produced, Australia
ABS (2017b) Australian Bureau of Statistics. Catalogue No. 71210DO001_201516 agricultural commodities, Australia- 2015–16. See Table 4: agricultural commodities–Queensland–2015–16
ABS (2017c) Australian Bureau of Statistics. Catalogue no 46180 water use on Australian farms 2015–16 released 7 July 2017
Australian Energy Regulator (2017) Queensland: small business customer disconnections. https://www.aer.gov.au/retail-markets/retail-statistics/queensland-small-business-customer-disconnections. Accessed 15 June 2017
Australian Farm Institute (2017) Australia exports enough food for 61,539,975 people – give or take a few! http://www.farminstitute.org.au/ag-forum/australia-exports-enough-food. Accessed 13 June 2017
Australian Government (2017a) Australian weather and seasons – a variety of climates. http://www.australia.gov.au/about-australia/australian-story/austn-weather-and-the-seasons. Accessed 1 July 2017
Australian Government (2017b) The basin plan 2012. Federal Register of Legislation. https://www.legislation.gov.au/Details/F2017C00078. Accessed 20 June 2017
Brooks S, Loevinsohn M (2011) Shaping agricultural innovation systems responsive to food insecurity and climate change. Nat Res Forum 35(3):185–200
Bureau of Meteorology (2017) Climate summaries archive. http://www.bom.gov.au/climate/current/statement_archives.shtml. Accessed 30 June 2017
Climate Council (2016) On the front line: climate change and rural communities. Climate Council of Australia Limited
Climate Change Council (2017) Hot and dry: Australia's Weird Winter. www.climatecouncil.org.au/2017-weirdwinter
CME (2012) Electricity prices in Australia: an international comparison. A report to the energy users association of Australia. Carbon and Energy Markets, March 2012

CSIRO (2012) Climate and water availability in south-eastern Australia: a synthesis of findings from phase 2 of the south eastern Australian climate initiative (SEACI). CSIRO, Australia

CSIRO and ENA (2017) Unlocking value: microGrids and stand alone systems. Roles and incentives for microgrids and stand alone power systems. A partnership between the Energy Networks Association and CSIRO

Davis G (2016) Earthquakes: do we have a credible disaster waste management strategy? Inside Waste. Issue 70: 24–25

Davis G, Chamberlin A (2016) Transitioning tariffs in agriculture. Elect Commun Data 15(5):42–44

Department of Agriculture (2017) Australian food statistics 2012–2013. www.agriculture.gov.au/Site CollecitonDocuments/ag-food/publicaitons/food-stats//australia-food-statistics-2012-13.pdf. Accessed 20 June 2017

Department of Agriculture and Fisheries (2016) Annual report 2015–16. Queensland Government, Brisbane

Department of Agriculture and Fisheries (2017) Agriculture and food research, development and extension: 10-year roadmap. June 2017

Department of Agriculture and Water Resources (2017a) Intergovernmental agreement on a national water initiative. http://www.agriculture.gov.au/water/policy/nwi. Accessed 24 May 2017

Department of Agriculture and Water Resources (2017b) Considering climate change and extreme weather events in water planning and management 2017: a module to support water planners and managers consider and incorporate possible impacts from climate change and extreme weather events on water resources. http://www.agriculture.gov.au/water/policy/nwi/climate-change. Accessed 23 June 2017

Department of Environment and Heritage Protection (2017) Link energy. https://www.ehp.qld.gov.au/management/linc-energy/. Accessed 1 July 2017

Department of Natural Resources and Mines (2017) Catchment areas. https://www.dnrm.qld.gov.au/water/catchments-planning/catchments. Accessed 23 June 2017

DIICCSRTE (2013) Department of industry, innovation, climate change, science, research and tertiary education. Climate adaptation outlook: a proposed national adaptation assessment framework, Australian Government, Canberra

Edwards B, Gray M, Hunter B (2015) The impact of drought on mental health in rural and regional Australia. Soc Indic Res 121(1):177–194

Garnaut R (2008) The Garnaut climate change review. Department of Climate Change, Canberra

Gell P, Reid M (2014) Assessing change in floodplain wetland condition in the Murray Darling Basin, Australia. Anthropocene 8:39–45

Hagger V, Fisher D, Schmidt S, Blomberg S (2013) Assessing the vulnerability of an assemblage of subtropical rainforest vertebrate species to climate change in south East Queensland. Austral Ecol 38:465–475

Hayman P, Rickards L, Eckard R, Lemerle D (2012) Climate change through the farming system lens: challenges and opportunities for farming in Australia. Crop Pasture Sci 63:203–214

Hulme M (2016) 1.5°C and climate research after the Paris agreement. Nat Clim Chang 6:222–224

IPCC (2012) Special report on managing the risks of extreme events and disasters to advance climate change adaptation (SREX). Intergovernmental panel on climate change. Cambridge University Press, Cambridge

Jackson TM, Khan S, Hafeez M (2010) A comparative analysis of water application and energy consumption at the irrigated field level. Agric Water Manag 97:1477–1485

Jackson TM, Hanjra MA, Khan S, Hafeez M (2011) Building a climate resilient farm: a risk based approach for understanding water, energy and emissions in irrigated agriculture. Agric Syst 104:729–745

Johnson O, Louise K (2017) Co-exploring the water-energy-food nexus: facilitating dialogue through participatory scenario building. Front Environ Sci 5:Article 24. https://doi.org/10.3389/fenvs.2017.00024

Keogh M (2011) Australia still at the bottom when it comes to farm subsidies. AgForum, 23 September. Australian Farm Institute, Sydney. www.farminstitute.org.au/ag-forum/australia_still_at_the_bottom_when_it_comes_to_farm_subsidies. Accessed 30 June 2017
Khan S, Rana T, Yuanlai C, Blackwell J (2006) Can irrigation be sustainable? Agric Water Manag 80:87–99
Khan S, Khan MA, Hanjra M, Mu J (2009) Pathways to reduce the environmental footprints of water and energy inputs in food production. Food Policy 34:141–149
Kiem AS, Austin EK (2013) Drought and the future of rural communities: opportunities and challenges for climate change adaptation in regional Victoria, Australia. Glob Environ Chang 23(5):1307–1316
Kiem AS, Askew LE, Sherval M, Verdon-Kidd D, Clifton C, Austin E, McGuirk PM, Berry H (2010) Drought and the future of rural communities: drought impacts and adaptation in regional Victoria, Australia. National Climate Change Adaptation Research Facility, Gold Coast
King AD, Karoly DJ, Henley BJ (2017) Australian climate extremes at 1.5 °C and 2 °C of global warming. Nat Clim Chang 7:412–416
McDonald J (2010) Climate change adaptation in South-East Queensland. Human settlements: issues and context. A report for the South East Queensland climate adaptation research initiative. Griffith University
McGovern M (2014) Subprime agriculture and Australia. Econ Anal Policy 44(3):243–258
MDBA (2014) Basin-wide environmental watering strategy 2014. Murray Darling Basin Authority 24 November 2014
Mitchell D, James R, Forster P, Betts RA, Shiogama H, Allen M (2016) Realising the impacts of a 1.5°C warmer world. Nat Clim Chang 6:735–737
Molden D, Oweis T, Steduto P, Bindraban P, Hanjra M, Kijne L (2010) Improving agricultural water productivity: between optimisation and caution. Agric Water Manag 97(4):528–535
Moshin M (2015) Agribusiness financing in Australia: issues and research agenda. Int J Econ Financ 7:1–18
Mullen JD (2010) Trends in investment in agricultural R&D in Australia and It's potential contribution to productivity. Aust Agribus Rev 8(4):18–29
Naish S, Mengersen K, Hu W, Tong S (2013) Forecasting the future risk of Barmah Forest virus disease under climate change scenarios in Queensland, Australia. PLoS One 8(5):e62843
National Centre for Engineering in Agriculture (2015) Improving Energy Efficiency on Irrigated Australian Cotton Farms Publication No. 1005371/1
Perry C, Stdeuto P, Allen RG, Burt CM (2009) Increasing productivity in irrigated agriculture: agronomic constraints and hydrological realities. Agric Water Manag 96:1517–1524
Queensland Competition Authority (2017) Queensland Government Gazette – Various. http://www.qca.org.au/Electricity/Regional-consumers/Reg-Electricity-Prices/Archive/Regulated-Electricity-Prices-2007-08. Accessed 24 July 2017
Queensland Farmers' Federation (2017) Energy savers program case studies: irrigation. http://www.qff.org.au/newsroom/case-studies/?in=1484. Accessed 3 May 2017
Queensland Government (2015) Queensland Government population projections, 2015 edition and ABS 3101.0, Australian demographic statistics. Queensland Government Statistician's Office
Queensland Government (2016) Business and industry portal. https://www.business.qld.gov.au/industry/agriculture/agriculture/plant-industries. Accessed 30 December 2016
Queensland Government (2017a) Statistics and facts: Queensland. https://www.qld.gov.au/about/about-queensland/statistics-facts/facts. Accessed 20 April 2017
Queensland Government (2017b) Queensland population counter. http://www.qgso.qld.gov.au/products/reports/pop-growth-qld/qld-pop-counter.php. Accessed 20 July 2017
Queensland Government (2017c) 2016 State of the Industry. Department of Tourism, Major Events, Small Business and the Commonwealth Games. Report available (30 June 2017) at https://publications.qld.gov.au/dataset/state-of-the-industry/resource/a434f571-c798-4c27-a894-ff456405f5a3. Accessed 2 July 2017

Queensland Government (2017d) Department of Natural Resources and Mines. Lower Burdekin Groundwater Strategy project: Discussion Paper. Unpublished

Queensland Productivity Commission (2016) Electricity pricing inquiry 2016. Chapter 10: rural and regional industries – transitional and obsolete tariffs

Qureshi M, Whitten S, Mainuddin M, Marvanek S, Elmahdi A (2013) A biophysical and economic model of agriculture and water in the Murray-Darling Basin, Australia. Environ Model Softw 41:98–106

Reisinger A Kitching R Chiew F Hughes L Newton P Schuster S Tait A Whetton P (2014) Climate change 2014: impacts, adaptation and vulnerability. Part B: regional aspects. Contribution of Working Group II to the Fifth Assessment Report of the Intergovernmental Panel on Climate Change. [Barros, V.R., C.B. Field, D.J. Dokken, M.D. Mastrandrea, KJ Mach, TE Bilir, M Chatterjee, KL Ebi, YO Estrada, RC Genova, B Girma, ES Kissel, AN Levy, S MacCracken, PR Mastrandrea, and LL White (eds.)]. Cambridge University Press, Cambridge, United Kingdom and New York, NY, USA. pp 1371–1438

Rickards L (2012) Critical breaking point? The effects of climate variability, climate change and other stressors on farm households. Final report to Birchip Cropping Group, Birchip, Victoria

SEQWater (2017) South East Queensland water dam levels. http://www.seqwater.com.au/water-supply/dam-levels. Accessed 20 June 2017

Sheng Y Mullen JD, Zhao S (2010) Has growth in productivity in Australian broadacre agriculture slowed? AARES Conference Paper, Adelaide. February 2010

Simshauser P (2017) Monopoly regulation, discontinuity and stranded assets. Energy Econ 66:384–398

Smith I, Skytus J, McAlpine C, Wong K (2013) Squeezing information from regional climate change projections – results from a synthesis of CMIP5 results for south-East Queensland, Australia. Clim Chang 121(4):609–619

SunWater (2017a) Water storage levels. http://www.sunwater.com.au/__data/win/reports/win_storages.htm. Accessed 20 June 2017

SunWater (2017b) Storage forecast models. http://www.sunwater.com.au/__data/storageforecast/Forecast_Portal.htm. Accessed 20 June 2017

TSBE (2017) Presentation by Warwick King, CEO Australian Pacific LNG The Toowoomba and Surat Basin Enterprise National Energy Summit, Toowoomba 12–13 July 2017

United Nations (2015) World population prospects: the 2015 revision, key findings and advance tables. Working Paper No. ESA/WP. 241. United Nations, New York

Vins H, Bell J, Saha S, Hess J (2015) The mental health outcomes of drought: a systematic review and causal process diagram. Int J Environ Res Public Health 12(10):13251–13275

Woodhead A, Sun X, Cotter J, Maraseni T (2015) Review of Asian consumers attitudes towards GM food and implications for agricultural technology development in Australia. Farm Policy J 12(3):37–44

Woodhead A, Earl G, Zhang S (2017) Integrating Australian agriculture with global value chains. Committee for Economic Development of Australia (CEDA), Melbourne

WWF (2015) World Wildlife Fund and Melbourne Sustainable Society Institute. Appetite for change: global warming impacts of food and farming regions in Australia WWF Australia

Zilberman D, Sproul T, Rajagopal D, Sexton S, Hellegers P (2008) Rising energy prices and the economics of water in agriculture. Water Policy 10:11–21

Chapter 6
Climate Change and Water Security Issues in Africa: Introducing Partnership Procurement for Sustainable Water Projects in Nigeria

Abdullahi Nafiu Zadawa and Abdelnaser Omran

Abstract In Nigeria, like in most African countries, the effects of climate change are posing a challenge to the sustainability of water security. Africa experiences more severe effects of climate change; scientific and statistical prediction shows that Africa is likely to experience an increase in temperature, rises in sea level, and changes in rainfall pattern. Nigeria as the most populous country in Africa is also highly vulnerable to the effects of climate change. The impact of climate changes affects the characteristics of freshwater resources in Nigeria, and some rivers and lakes were observed with a reduction in flow rate and networks due to a decrease in rainfall and higher evaporation. These posed a unique challenge to the water security sector in Nigeria, and a more innovative urban water management strategies and policies are urgently needed to safeguard people's life and enhance economic condition in African countries, especially in Nigeria. As a projected remedy, the study proposes a conceptual framework which outlines the potentials of partnership procurement approach for the development and management of water supply in Nigeria. It is a more adaptive project strategy that is all-inclusive and innovative which appropriately guides procurement professional against risk and uncertainties especially those caused by unpredictable climate change.

Keywords Water security · Climate change · Partnership · Procurement · Sustainable · Nigeria

A. N. Zadawa (✉)
Quantity Surveying Department, School of Environmental Technology, Abubakar Tafawa Balewa University, Bauchi, Nigeria
e-mail: nzabdullahi@atbu.edu.ng

A. Omran
School of Economics, Finance & Banking, Universiti Utara Malaysia, Sintok, Kedah, Malaysia

A. Omran, O. Schwarz-Herion (eds.), *The Impact of Climate Change on Our Life*, https://doi.org/10.1007/978-981-10-7748-7_6

6.1 Introduction

The concept of water security aims at instigating a scheme to provide secured and sustainable water to the populace, thus ensuring control of water as a commodity among the household (Narcisse 2010). Water security is regarded as the key solution to water crises in the twenty-first century; the concept was launched at the Ministerial Declaration of The Hague in March 2000. Accordingly, water security entails providing people and community with adequate, reliable, and safe water for their different needs; people and community are able to utilize all the opportunities offered on water resources and are safeguarded against water-related diseases. Water as a commodity comes from many sources including ground- and surface water. In most Nigerian cities, water for domestic use comes from public water resources, where there is scarcity conflict. Therefore, the idea of water security means to have a scheme in place to ensure scarcity does not cause conflict. Access to clean and safe water for household use is essential for human development. Yet, approximately 884 million people most of whom live in developing countries lacked access to improved water resources as reported by the UNDP in 2008.

Considering the role of water in terms of human welfare and economic development, water resource management is vital to Nigeria's effort in poverty reduction, improving health condition and ensuring food security. Despite so much effort to enhance water supply development in Nigeria, still over 43% of country's population lack access to safe water supply (Gbadegesin and Olorunfemi 2007). From the broader perspective, it is estimated that between 75 and 250 million, African population will be prone to water security issues, food, and environmental stress due to climate change by the year 2022 (IPCC 2007). For the fact that Africa seems to be affected more by the effects of climate change, Nigeria as the most populous country on the continent is more vulnerable to such effects. At the moment, certain ecological problems that have already surfaced in Nigeria have been a major implication of climate change on the environment. Also, there are other obvious factors of climate change with negative impact on water and wetland resources in Nigeria (Ebele and Emodi 2016).

Nigeria highly depends on freshwaters, and the impact of climate changes affects the characteristics of freshwater resources. Some rivers and lakes were observed with a reduction in flow rate and networks due to a decrease in rainfall and higher evaporation (Ebele and Emodi 2016). For instance, Sokoto River in northwestern Nigeria was observed with decrease in flow which causes a water shortage for urban supply, agricultural and hydropower uses (Nwankwoala 2012). Continuous climate change is likely to have substantial effects on water supply, for instance, reduction in water, and availability is more expected in mid-latitudes water-stressed areas, while at high latitudes, water supply is expected to increase. A number of studies that examine the effects of climate change on groundwater in Africa include the studies of Doll and Fiedler (2008) that predicted larger percentage decrease in groundwater recharge than runoff water in southwestern Africa. A study conducted by Cave et al. (2003) in southern Africa explained how a decrease in annual rainfall significantly affects groundwater recharge in the region. Most of these studies were obviously silent on the role partnership procurement approach for sustainable water projects to adequately mitigate the water security issues due to climate changes; this study aims to filling the identified gap.

6.2 Climate Change and Water Security Issues

Climate change is one of the major threats to the environment generally in the twenty-first century. Its impact is undeniably causing a serious threat to the elements of the environment especially water resources which necessitate the need to explore adaptive mitigation measures (Ebele and Emodi 2016). Recently, climate change and its impacts on water resources have become an issue of global concern (Misra 2014). As a result, a number of studies were conducted searching for adaptation of new planning process, new policies, feasible frameworks, and more innovative technological approaches (Adger et al. 2005). But, most of these policies were not properly implemented. The impact of climate changes on water resources is increasing day by day globally posing a challenge to the prime goal of water security concept. About 3 billion people rely on groundwater for consumption and other purposes, and global climate models predict significant changes in the amount of rainfall and air temperature which affects groundwater recharge subsequently (Kurylyk and MacQuarrie 2013). According to IPCC report (2008), predicted climate changes significantly affect the pattern of annual rainfall, river flows, and sea level across the globe. Earlier in 1998, the WHO reported that water security issues are severe in developing countries, where approximately about 1.2 billion people in more than 20 developing countries are faced with water security issues. It is also predicted that more than 30 countries in Africa and Asia will be affected by the year 2020 (Asare 2004).

6.3 The Impact of Climate Change on Water Resources in Africa

Africa experiences more severe effects of climate change; scientific and statistical prediction shows that Africa is likely to experience an increase in temperature, rises in sea level, and changes in rainfall pattern. Specifically, access to water may be the single biggest impact of climate variability in Africa in the near future as reported by the United Nations. In sub-Saharan Africa, the amount of rainfall is predicted to drop by 10% by the year 2050, and this will cause a shortage of groundwater. There has been an increase in temperature in the most part of the African continent since the 1960s which indicates greater warming (Boko et al. 2007). Since the last three decades, the warming seems to be consistent over the continent, and it has been increasing at the rate of 0.5 degree Celsius per decade (Hulme et al. 2001). Most importantly, the most noticeable change in climate in most of Africa has been the long-term reduction in rainfall especially in western Africa and the Sahel part (Nicholson and Grist 2001). The statistical record indicates that a decrease in the amount of rainfall in Africa has been 2.4 ± 1.3 per decade averagely, while western Africa was observed with the fastest decreasing rate estimated at −4.2 ± 1.2 per decade (Hulme et al. 2001). Bates et al. (2008) posited that African population that might be at risk of water stress and scarcity could increase from 2000 to 2025. This prediction is directly related to increasing in population, but it will further complicate water conflict (UNECA-ACPC 2011).

The future climate change in Africa on water resources is anticipated in two aspects: first is the unpredictability of rainfall variation (UNECA-ACPC 2011), and second is the sensitive nature of runway water in relation to evapotranspiration (Elshamy et al. 2009). Introducing new strategies that will provide future precautions against such impact becomes necessary since most of the policies regulating water supply system, management and operating procedure in Africa do not take effect of climate change into consideration (ACPC 2011). Introducing partnership procurement approaches for water supply projects is required in order to cope with the impacts of climate change on water in most African countries as introduced by the current study.

6.4 The Scenario in the Nigerian Context

Countries that fall within the developed economies face less severe effects of climate changes due to advance adaptation strategies and high financial, economic status. In Nigeria and most other developing countries, the effect of climate change is particularly severe due to lack of adaptation strategies and poor financial and economic capacities. Climate change in Nigeria mostly affects the frequency of rainfall distribution, increase in sea level, and rising number of a heat wave (Urama and Ozor 2010). The frequent unusual weather changes that affect rainfall distribution in different parts of Nigeria are the most obvious negative effects of climate change (Urama and Ozor 2010). For instance, low-lying coastline areas experience high frequency of coastal erosion and flooding. The decrease in rainfall is experienced mostly in Northern Nigeria making the region desert-prone and causing more desertification. Some of the major signs of the impact of climate change in Nigeria as mentioned by Victor Fodeke in an interview with Daily Trust Newspaper is the rain being experienced when it is not expected in most part of the country, as well as the large negative effects on agricultural produce (Yahaya 2009). It is well known that southern part of Nigeria experienced higher rainfall compared with the northern region; due to the impact of climate change, the region is now confronted by irregular rainfall pattern. Whereas, desert encroachment is one of the major threats to the northern region which adversely decreases the amount of surface water (Obioha 2008). In the northern region, most of the water supply projects especially the urban water supply system are sourced from rivers. Originally, urban water supply by the state waterboards in Nigeria was initiated by the colonial government. The Colonial administration initiated and established an urban domestic water supply to improve environmental sanitation, the level of personal hygiene, and health condition of the populace throughout the country. Due to the lack of technical capacity to effectively operate and maintain these water projects by the postcolonial governments, coupled with continual negative impacts of climate change on the water resources, the public waterboards in most Nigerian urban settlements are no longer fully functional. Supply of water has been on the steady decline both in terms of quality and distribution; these further resulted in poor sanitation and hygiene as well as the spread of waterborne diseases (Amori et al. 2012). Moreover, treatment plant and pumping machines require a constant and steady power supply with is currently not feasible in Nigeria considering the acute shortage of power supply. This has added to the impacts of climate changes on the water security in Nigeria, which obviously is beyond the technical and financial capacity of the government alone. Therefore, the current study proposes the introduction of partnership procurement strategies between private and public sector entities for water supply projects given the technical and financial potentials associated with the private sector.

6.5 Hypothesized Partnership Procurement Framework for Water Supply in Nigeria

Partnership for water supply projects is one of the strategies introduced by the World Bank known as Water Partnership Program (WPP). The WPP is a collaborative initiative between the World Bank and the Governments of the Netherlands, the UK, Australia, and Denmark. It was introduced to reduce the level of poverty and to strengthen climate change resilient efforts in water resource management. The WPP enables the World Bank to sufficiently respond to country's water demand, thus introducing an innovation in the water sector, in addition to the change in global policy dialogue. Phase I of the WPP has successfully ended with good records, and by the end of June 2012, the World Bank has launched phase II of WPP with wider coverage to significantly assist countries to overcome the effects of climate change on water supply with a better alternative for water management and services. Based on this concept and other associated collaborative model for water supply projects, this study proposes a conceptual framework (Fig. 6.1) for partnership procurement of water supply project in Nigeria to adequately respond to the effects of climate changes on water security issues.

The proposed model is based on the concept of mediation. In recent times, mediation for data analysis in behavioral and applied science fields including procurement management is used in analyzing simple and complex relationships (Hayes 2012). Mediation as a modern statistic with different approaches allows researchers to explore and understand how and why relationships and/or effects exist between study variables (Hayes 2012; Wu and Zumbo 2008). Mediation is said to occur in a given framework when the effects of the relationship between independent con-

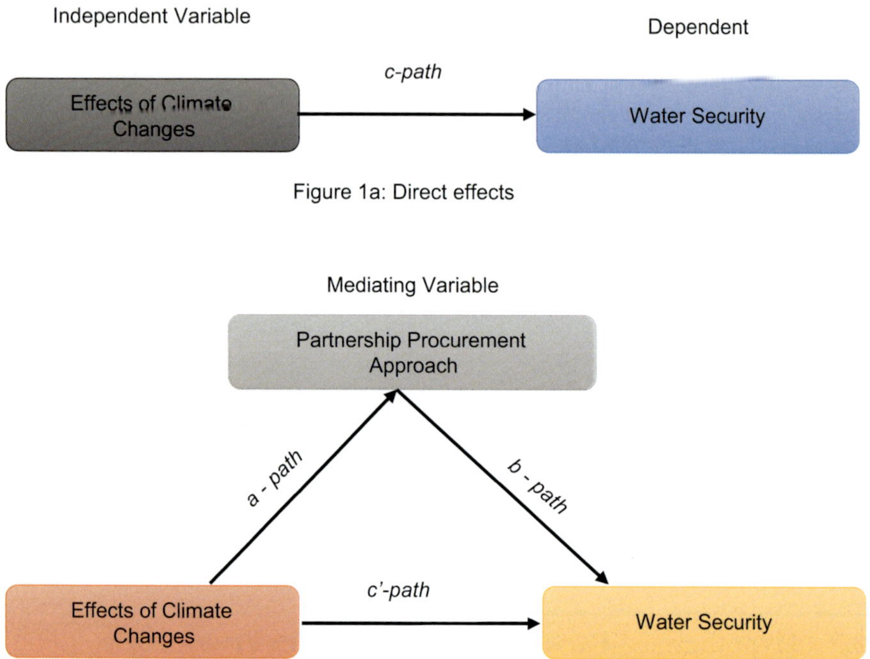

Fig. 6.1 Conceptual framework of the study

struct and the outcome variable are mediated by an intermediary variable introduced as a mediator of the framework (Hayes et al. 2011). The framework of the current study as presented in Fig. 6.1 has two major components, i.e., (i) direct effect model (Fig. 6.1a) which shows the direct relationship between effects of climate change (*independent variable*) and water security (*dependent variable*) – *"c-path"* denotes the coefficient for the effect of the relationship in the direct effect model – and (ii) indirect effect model (Fig. 6.1b) which shows the effects of climate change (*independent variable*) on water security (*dependent variable*) through the mediating variable (*partnership procurement approaches*). For the indirect effect model, "a-path" and "b-path" denote indirect effect coefficients, while *c'*- the path is the coefficient of the direct effects after adding the mediating variable.

In view of the obvious effects of climate change on water security coupled with the insufficient public budget and technical capabilities, the introduction of partnership procurement in water projects has become a priority (Golooba-Mutebi 2012). Although water projects developed through partnership procurement are characterized with challenges such as appropriate water technologies, high transaction costs, and regulatory weakness, yet there exists evidence of effective performance in such projects compared to public-owned waterboards (Gopakumar 2010). Partnership for water supply projects has made significant long-lasting solution toward delivering water supply projects. According to UNDP (2006), over 160 million people were being supplied with water by private water operators as of 2007, and approximately over 24 million people were provided with water access through partnership procurement strategies.

6.6 Conclusion and Recommendation

Climate change is posing uncertainty to water security in Africa, in Nigeria; the high level of uncertainty posed by climate variability to water security sector is quite challenging and requires a more innovative adaptation as well as resilient approaches. The need for more accurate predictive measures against climate change is also necessary. Considering the economic status of Nigeria as a developing country, this study concludes that the government alone cannot take up the challenge. Therefore, the need for cooperation and collaboration with the private sector and nongovernmental organization to develop more innovative strategies for better adaptation and development of more sustainable water supply projects. The study recommends for the introduction of partnership procurement strategies for water supply projects in Nigeria to adequately tackle impacts of climate changes on water resources. The government should develop partnership action plans as a mechanism to adequately plan and establish water projects which will incorporate proactive measures against impacts of climate changes.

References

ACPC (2011) Consultation workshop on detailing work plans of climate science, data and information, climate change and water, and climate change and agriculture. United Nation Economic Commission for Africa, African Climate Policy Centre ACPC, Addis Ababa

Adger WN, Arnell NW, Tompkins EL (2005) Successful adaptation to climate change across scales. Glob Environ Chang 15(2):77–86

Amori AA, Eruola AO, Makinde AA, Ufoegbune GC (2012) Assessment of problems affecting public water distribution in a humid tropical zone. Int J Hydraul Eng 1(3):21–24

Asare YBO (2004) Household water security and water demand in the Volta Basin of Ghana. Ph.D thesis, University of Accra, Ghana

Bates BC, Kundzewicz ZW, Wu S, Palutikof JP (eds) (2008) Climate change and water, Technical paper of the intergovernmental panel on climate change. IPCC Secretariat, Geneva, 210 pp

Boko M, Niang I, Nyong C, Vogel A, Githeko M, Medany B, Osman-Elasha, Tabo R, Yanda P (2007) Africa climate change 2007: impacts, adaptation and vulnerability. In: Parry ML, Canziani O, Palutikof JP, Van Der Linden PJ, Honson CE (eds) Contribution of working group II to the fourth assessment report of the intergovernmental panel on climate change. Cambridge University Press UK, Cambridge, pp 433–467

Cavé L, Beekman HE, Weaver J (2003) Impact of climate change on groundwater resources estimation. In: Xu Y, Beekman HE (eds) Groundwater recharge estimation in Southern Africa. UNESCO IHP Series No. 64

Doll P, Fiedler K (2008) Global-scale modelling of ground water recharge. Hydrol Earth Sys Sci Discuss 12(3):863–885

Ebele NE, Emodi NV (2016) Climate change and its impact on Nigerian economy. J Sc Res Rep 10(6):1–13

Elshamy ME, Seierstad IA, Sorteberg A (2009) Impacts of climate change on Blue Nile flows using bias-corrected GCM scenarios. Hydrol Earth Syst Sci 13(5):551–565

Gbadegesin N, Olorunfemi F (2007) Assessment of rural water supply management in selected rural areas of Oyo State, Nigeria, Working paper series no. 49. African Technology Policy Studies Network, Nairobi

Golooba-Mutebi F (2012) In Search of the right formula: public, private and community-driven provision of safe water in Rwanda and Uganda. Public Adm Dev 32(4–5):430–443

Gopakumar G (2010) Transforming water supply regimes in India: do public-private partnerships have a role to play? Water Altern 3(3):492–511

Hayes AF (2012) Process: a versatile computational tool for observed variable mediation, moderation, and conditional process modeling [White paper] Retrieved on December 12, 2016, from http://imaging.mrc-cbu.cam.ac.UK/statswiki/FAQ/SobelTest

Hayes AF, Preacher KJ, Myers TA (2011) Mediation and the estimation of indirect effects in political communication research. In: Bucy EP, Holbert RL (eds) Sourcebook for political communication research: methods, measures, and analytical techniques, vol 23. New York, Routledge, pp 434–465

Hulme M, Doherty R, Ngara T, New M, Lister D (2001) African climate change: 1900–2100. Clim Res 17(2):145–168

Intergovernmental Panel on Climate Change (IPCC) (2007) Climate change impact adaptation and vulnerability: contribution of working group II to the fourth assessment report of the IPCC. Cambridge University Press, Cambridge

Intergovernmental Panel on Climate Change (IPCC) (2008) Climate change and water, Intergovernmental panel on climate changes technical report IV, June 2008

Kurylyk BL, MacQuarrie KT (2013) The uncertainty associated with estimating future groundwater recharge: a summary of recent research and an example from a small unconfined aquifer in a northern humid-continental climate. J Hydrol 492:244–253

Misra AK (2014) Climate change and challenges of water and food security. Int J Sustain Built Environ 3(1):153–165

Narcisse KA (2010) Water security and the poor: evidence from rural areas in Cote d'Ivoire. Natural resource management and economic development research paper. University of Cocody-Abidjan, Abidjan, Cote D'Ivoire

Nicholson SE, Grist JP (2001) A conceptual model for understanding rainfall variability in the West African Sahel on interannual and interdecadal timescales. Int J Climatol 21(14):1733–1757

Obioha E (2008) Climate change, population drift, and violent conflict over land resources in North Eastern Nigeria. J Hum Ecol 23(4):311–324

UNDP (2006) Human development report 2006, Beyond scarcity: power, poverty and the global water crisis. United Nations Development Programme (UNDP), New York. http://hdr.undp.org/hdr2006/

UNECA-ACPC (2011). Climate Change and Water in Africa: Analysis of Knowledge Gaps and Needs. United Nations Economic Commission for Africa African Climate Policy Centre. Working Paper 4.

Urama KC, Ozor N (2010) Impacts of climate change on water resources in Africa: the role of adaptation. African Technol Policy Stud Netw 29(1):1–29

Wankwoala HO (2012) Case study on coastal wetlands and water resources in Nigeria. Eur J Sustain Dev 1(2):113–126

Wu AD, Zumbo BD (2008) Understanding and using mediators and moderators. Soc Indic Res 87(3):367–392

Yahaya NA (2009) Unusual rain, evidence of climate change. All Africa.com. Retrieved on April 25th, 2017 from http://allafrica.com/stories/200901220158.html

Chapter 7
Waste Management and the Need for a Better Approach on Global Sustainability

Florina Bran, Carmen Rădulescu, and Alexandru Bodislav

Abstract Waste represents in European countries as one of the biggest issues regarding environmental protection. The population consumes immense quantities of natural resources, renewable and depleted, and also tries to value environmental factors that if consumed they will damage the environment on the long run. Managing waste represents and targets all activities needed to value, eliminate, transport, use, and collect waste. To create optimal waste management, we must target reducing consumption of natural resources. With population growth, we also have a new measure targeted, especially because there are limited natural resources that are consumed more than supported to restock. The main objectives of managing waste are environmental protection and healthcare status and also keeping natural resources and their extractive environment intact. Our chapter offers a new approach on waste and the issues it develops, how waste will complicate our daily actions and influence the decay of the environment. To increase the value and ease the process of manufacturing, we need renewable materials to be sorted and separated from other types of products; this way they cannot cross-contaminate each other. Deposits of urban and industrial waste have a huge impact on the environment because they pollute the air, the soil, and all types of waters, and they modify on the spot, but for the long run, the soil's fertility, the environmental long-term output, and the aesthetics of the environment. We should emphasize the fact that having the necessary information and realizing the dissemination to the population about the importance of an optimal waste management will help future projects on managing waste, and it will give a push to ignite solutions for environmental issues, protecting green areas and eliminating the risks for human health, especially because we also need to protect the communitarian sentiment at all costs.

Keywords Sustainable · Waste management · Development · Environmental protection

F. Bran · C. Rădulescu · A. Bodislav (✉)
The Bucharest University of Economic Studies, București, Romania
e-mail: dumitru.bodislav@economie.ase.ro

A. Omran, O. Schwarz-Herion (eds.), *The Impact of Climate Change on Our Life*, https://doi.org/10.1007/978-981-10-7748-7_7

7.1 Introduction

In the context of sustainable development, efficient waste management involves taking into account key issues related to waste and residue reduction, possibly recovering the value they have, taking into account the environmental protection restrictions and the economic restrictions. The goal of waste management is to maximize the conservation of nonrenewable resources and minimize negative environmental impacts of waste.

Sustainable development is a strategy by which communities seek ways of economic development benefiting from the local environment or reducing the benefits of quality of life. It has become an important guide for communities who have noticed that traditional modes of planning and development create more than solving environmental problems. Overall, sustainable development can be achieved by building on the following ideas:

1. Raw materials must be administered so as to facilitate and stimulate optimal recycling and the use of waste products.
2. Improving energy consumption and production based on available resources.
3. A new vision on lifestyle and consumption in society.

There is currently a quantitative worsening of pollution and also a change of ecological problems. As remarkable phenomena, repeated local pollution in thousands and thousands of situations leads to global manifestations, including the intensification of the greenhouse effect, which is a conclusive example. It can be noticed that the geographical origin of the pollution is "disconnected" from its place of manifestation, and some environmental problems will easily affect the planet (Rojanschi et al. 2004).

Efficient waste management is based on some specific aspects that are linked with reducing waste and residual output; this way we target recovering valuable natural resources especially by creating restrictions that protect the environment and encourage economic growth. The main target of waste management is maximizing the preservation of unrenewable and renewable natural resources but also minimizing the negative effects of waste on the environment. Human kind's development represents a pressure point on the environment that could be extremely dangerous on the long run if not regulated. The European Commission is preoccupied on creating and keeping a clean environment, and because of this, there were developed industrial practices that could assure a better standard of living with a small impact on the environment, also for today's population but especially for future generations. Starting from the year 2001, durable development strategy was developed. The strategic instruments developed were targeted toward durable and sustainable production and optimal and rational consumption. These concepts were used to promote a longer life span of consumer goods; by this way that good would not evolve as waste and will be used for a longer period of time and to monitor the effects of using resources on the environment. The universal strategy is based only on reducing the quantity of waste and increasing the flow of recyclable materials and reducing energy consumption. Better manufacturing and optimal supply for the population are directly linked with waste production, and because of this, we also need a better approach on how we develop boxes for our goods, recyclable materials being the keyword in the process.

7.2 Waste Classification

Wastes are those substances or debris which the holder discards and has the intention or obligation to dispose them. Waste is generally the last step in the life cycle of a product (the time interval between the date of manufacture and the date when the product becomes waste). Waste is divided in terms of nature and production sites (Rojanschi et al. 2002, p. 258):

(a) Household wastes – are the result of household or consumer activities. Municipal waste, worldwide, exceeds 18 million tons a day, the authorities being responsible for the costs of all this waste production. According to the extent to which the authorities solve the waste problems, the assessment of the economic-cultural growth index can also be made, and the waste can be taken over with the current pre-collection or collection systems in the localities.
(b) Street wastes – are the result of the daily life of the population, from animals, from the accumulation of solid substances in the atmosphere, and from green spaces.
(c) Wastes that are assimilated to household waste – come from both the administrative sector and the public sector, as well as trade and industry.
(d) Bulky waste – because of its size, solid waste cannot be taken over but needs a differentiated treatment.
(e) Construction or demolition waste.
(f) Hazardous waste – is the waste that is hazardous to the health of the population and the environment. They are the result of bacteria, paints, drugs, organic chemicals, and flammable and non-flammable explosives, adhesives, and other chemicals that can threaten people's health. Due to incorrectly positioned dumps, it has been shown that they also contain toxic composition.
(g) Agricultural waste – is derived from livestock manure and from meat and animal slaughterhouses. These are composed of organic waste that needs to undergo a collection process and needs its own treatments, including animal manure from crops and forests, as well as residues from the food industry.
(h) Hazardous and industrial wastes – include various types of waste resulting from food, tobacco products, wood products, and paper products; from farmers, organizations, and entrepreneurs; and even from laboratories. Industrial waste is about 40 times higher than waste from municipalities, regions, or manufacturing industries. These require a controlled collection. Hazardous waste is treated and sorted according to the final treatment options and the main characteristics.

(i) Hospital wastes – are wastes from sanitary units and hospitals and infectious disease patients, which are very dangerous due to the increased risk of infection. Special hospital waste should be packaged carefully to avoid contamination and treated separately in a high-temperature combustion chamber so that all organic matter is completely destroyed. Another method of treating hospital waste is a burial.

Waste generation is subordinated to the needs and habits of the consumer, being generated by technological processes and living conditions. The correct measurement process for waste generation is weighing platforms, recycling stations, and treating facilities with quality records according to type of waste, volume, weight, transport, etc. With these, estimates and forecasts can be made. If these platforms are missing, estimation of waste generation can be done through the per capita/employee/waste unit, by assessing the annual growth of the storage volume of the platform but also by the volume and frequency of collection of the containers. There are differences from country to country in terms of characteristics and quantities of waste, depending on technological, industrial, living, demographic, and civilization (Bold and Mărăcineanu 2004, p.16). Community guidelines in recent years, through the revision of the Waste Framework Directives and through the adoption of the thematic strategies on the prevention of waste generation and the sustainable use of natural resources, demonstrate the determination to favor the options on the top hierarchy of their management. Waste management, from the best to the least good for the environment, is as follows: prevention, reuse, recycling, energy recovery, and disposal through incineration or storage.

7.3 Collection, Transport, and Storage of Waste

The collection of waste is makeup, processing, and transport to the recovery or their neutralization, representing the most important in waste management. Collecting is performed according to the transmission system, being a negligible undeveloped. At present, the provision of domestic waste collection points is unsatisfactory; containers have a high degree of wear and, in many cases, do not have sufficient storage capacity making it impossible to selectively collect waste. Certain collection points do not have concrete platforms, so the quality of soil and groundwater is affected (National Agency for Environmental Protection). The collection and transport of solid waste is in most cases (especially the rural environment) with outdated equipment that creates discomfort, resulting in ugly smells and waste of waste during transport.

Collection containers are considered to be accessories closely related to transport vehicles, making it much more correct for them to be as close as possible to the transporting organ and their maintenance to be ensured by the public sanitation unit. The main problems in maintaining clean containers, seen as a task of real estate management, are (Rojanschi et al. 2002, p. 274):

1. Garbage containers are seen as an attachment to transport vehicles, being manufactured with different volumes according to standards. The containers must be made of durable materials so as to withstand the blows, have easy-to-use caps, and have tight sealed cover, to empty them quickly and easily. Their own weight is as small as one person to handle their transport and cleaning.
2. Storage spaces - specially compiled rooms for the placement and storage of garbage containers and the physical location of the facility where vehicles could be approached near to the storage locations of the provided containers. Provision must be made for the space required for the containers to be laid, both for cleaning and for access to the rest of the installation.
3. Municipal waste collection and transport activities in the region are organized differently according to size of the locality, number of people served, endowment, and form of ownership.

Collection of municipal waste is the responsibility of municipalities, *directly*, and the duty of the local councils, *indirectly*, by leasing service companies specializing in sanitation. Municipal waste from dwellings, institutions, and economic entities are preselected in containers located in specially designed spaces with different capacities.

Sanitation companies also deal with the collection and transport of street waste. Street cleaning is maintained with machines equipped with vacuum cleaners and road brooms, but manual collection is the most common method of collection.

The biggest problem of waste collection is the insufficient number of containers or their lack of trash cans, both at the outskirts of towns, localities and in crowded places in the city center.

Trash cans of 35–50 liters are rarely used today because they are uncomfortable from the transport point of view and in terms of available capacity, and bulky waste is nonslip and must be transported separately or stored under the dumps.

Collection methods:

1. Mixed collection – this is the simplest method, which does not involve large efforts of the waste generator, from the point of view of selection by type of waste. For the sorting process of mixed recyclable packaging, the mechanical sorting system is required, being sorted in different stages, with the help of suitable machines or by hand.
2. Selective collection – the collection intervals of mixed waste and recyclable materials must match the collection system used. Hygiene conditions can make the intervals between the gradual collection of mixed waste to be reduced by reducing the volume of waste by simultaneously introducing recyclable materials. Reinventing recyclable waste materials in the production process as a secondary raw material and taking over is an objective in terms of saving raw material but also reducing the quantities of scrapped waste.
3. Collection procedures – this can be done by emptying the bin, by changing the bin, and by collecting it in disposable bags.

Selective collection is determined by the recovery of reused materials much easier, prior to collection of urban waste or after it is collected in treatment plants. The agricultural use of waste is easier in the case of industrial treatment by composting (Ionescu and Lefter 2013). For the selective collection of the quantities of packaging waste from a certain type of material from both household and waste that could be absorbed, it is first necessary to identify sources of generation (population, services, commerce, industry).

The way in which selective collection methods should be implemented and established on two or more types of packaging materials shall be carried out in such a way as to achieve the targets set in the plan after the characteristic circumstance of the county. Waste shipment represents all processes that have as their starting point the end of the collection process and end with the disposal of waste at the disposal, treatment, and recycling facilities (ICIM Bucharest 2005). The main required sanitary requirements for the construction of garbage trucks are (Rojanschi et al. 2002, p. 276):

- Ensuring the loading of garbage without scattering or dust and noise emitting in a rapid way and ensuring a rapid discharge and transport of completely closed garbage.
- Equipping with tools to continuously advance garbage loaded.
- The vehicles are in line with the provisions on road traffic and traffic safety.
- Construction to be simple and safe to operate, safe extraction without disturbance.
- Taking into account the frequent stops and starts, ensuring that they have a safe starting and braking system.

In order to choose a vehicle, one must take into account an important factor, namely, the population density and the ratio of housing to several families, which determines the number of vehicle stops. Fundamental requirements for the performance of transport and collection waste (Rojanschi et al. 2002, p. 276):

1. Switching
   Operation – the vehicles must have the necessary equipment for an efficient, reliable, high-efficiency load and must have the brushes and rollers that help to efficiently sweep the streets and public roads. Vehicles must have the greatest possible waste collection capacity, be able to move at a speed and in safety, and discharge the recyclable materials as quickly as possible.
2. Dangers of reduced damage
   Operator safety – use of machines to ensure safety during work, to avoid high effort with controls, and to have optimum visibility from the control station and appropriate lighting at night.
   Environmental protection – avoid odor pollution or loss of waste during transport, make no mistakes, ensure stability in order not to overturn the vehicle, and have filters against harmful gas emissions and dust.

According to the way environmental protection is ensured, storage can be:

1. Simple storage that understands simple, unorganized discharging, not taking into account special environmental protection measures, being the most widespread system due to the fact that it was cheaper and more convenient but considered totally dangerous for public hygiene, emitting unpleasant odors.
2. Controlled storage is the main storage system, respecting hygiene and environmental protection conditions, and unloading is done on ramps and on special grounds.

Storage can be made depending on how the waste is collected:

1. The storage of waste presorted during collection is done within a time frame in which the final sorting is carried out, following the transport of the waste on the units that insert it into the production circuit. The storage platforms are located in areas outside the localities.
2. Storage of unsorted waste is based on the subsequent processing.
3. Surface platforms are used for food waste processed by composting, which is mixed with earth, shredded and fermented, and then handed over as fertilizer for agriculture.

Land plots and landfill must be constructed in such a way that pollution or contamination of groundwater or surface water does not occur, and if they are used as surfaces of dwellings or buildings, corrosion, slip, or compression effects occur.

7.4 Addressing Waste at European Level

At EU level, a number of documents have been implemented that are in line with waste management legislation. Thus, the first act is *Directive 2008/98 on waste and repealing certain Directives*. In Romania, this directive was implemented by *Law no. 211/2011 on waste regime*.

Depending on the type of waste, both at European and national level, the legislation includes several documents, hence *Directive 94/62 on packaging and packaging waste*, namely, *the Order 794/2012 of the Minister of Environment and Forests on the procedure for reporting data on packaging and packaging waste*.

Where waste is based on electrical and electronic equipment, at European level *Commission Directive 2005/369/EC laying down rules for monitoring compliance of Member States and establishing data formats for the purposes of Directive 2002/96/EC on waste electrical and electronic* and nationally *Order no. 1494/846/2016 of July 20, 2016 for the approval of the procedure and criteria for granting the operating license, revision, annual sighting and cancellation of the operating license of the collective organizations and approval of the operating plan for the producers who fulfill their individual obligations the licensing of authorized representatives and the composition and powers of authorization for management of waste electrical and electronic equipment.*

For waste from batteries and accumulators, at European level, the *Directive 2006/66 on batteries and accumulators and waste batteries and accumulators* was adopted. For this type of waste, the *Regulation 1103/2010 establishing, pursuant to Directive 2066/66/EC of the European Parliament and of the Council of capacity labeling rules for portable batteries and accumulators secondary* was also adopted.

In 2006, the *Directive 2006/21/EC on the management of waste from extractive industries* was approved. However, nationally, the *Government Decision no. 856/2008 on the management of waste from extractive industries* was enforced. For end-of-life vehicles, *Directive 2000/53 on end-of-life vehicles* was adopted (this directive was updated by the European Committee for Economy and Social in 2017).

An important objective of European waste policy is the safe disposal of waste that cannot be recycled and reused. To this end, the *Directive 96/59 on the disposal of polychlorinated biphenyls and terphenyls (PCB and PCT)* was implemented. Romania's provisions on this directive was done through *Government Decision no. 173/2000 regulating the special regime for the management and control of polychlorinated biphenyls and other similar compounds*. Also in order to eliminate waste that cannot be recycled and reused, the *Directive 75/439/EEC on waste oils* was established. Subsequently, it was repealed, and the provisions on the disposal of waste oils were included in *Directive 2008/98*/EC on waste. In addition to the European regulations in Romania, the *Government Decision no. 235/2007 on the management of waste oils* was approved.

As the sludge should be used under conditions which ensure the protection of soil and groundwater and surface water, *Directive 86/278 on the protection of the environment, especially the soil, when sewage sludge is used in agriculture* has been adopted at European level. In Romania, it was completed by *2004 344/708 Common Order of the Minister of Environment and Water and the Minister of Agriculture, Forests and Rural Development approving the technical rules on environmental protection in particular soil, when sewage sludge is used in agriculture*. In order for Romania to be able to capitalize on its development potential in key sectors, such as transport or waste management, it is necessary to adopt measures that will lead to an efficient use of European Union funds as well as an increase in the absorption rate. Managerial deficits and frequent changes in priorities have affected the quality of public investment.

At European Union level, Romania has agreed to meet certain environmental standards. Thus, enforcement measures are empowered so that national and EU funds are geared toward achieving these environmental standards. As a consequence, the first version of the National Waste Management Plan was made available to the public. With regard to biofuels, it is considered that they should not come from agricultural products. Also, the development of biofuels should be done without affecting food production rather from residual products, by-products, and even waste, such as forestry.

Pollution and global warming can be influenced both by how the products are produced and the services are offered and by how they are used. Thus, the non-use of high raw materials can lead to the depletion of natural resources and to the

Table 7.1 A comparison on generated waste between countries that are members of the European Union

Country/year	2004	2015
EU (28)	512	476
Austria	574	560
Belgium	485	418
Bulgaria	599	419
Czech Republic	279	316
Cyprus	684	638
Croatia	304	393
Denmark	695	789
Estonia	445	359
Finland	469	500
France	519	501
Germania	587	625
Greece	436	485
Ireland	737	587 (data for 2012)
Italy	540	486
Latvia	318	404
Lithuania	373	448
Luxemburg	679	625
Malta	623	624
Great Britain	602	485
Holland	599	523
Poland	256	286
Portugal	445	453 (data for 2014)
Romania	349	247
Slovakia	261	329
Slovenia	485	449
Spain	600	434
Sweden	460	447
Hungary	454	377

Source: Data compilation done by the authors by using EUROSTAT

generation of large quantities of waste. A great deal of waste generated by households is a municipal waste. But, they can also include similar waste generated by small businesses and institutions. Depending on the local waste management system, this part of municipal waste may vary from city to city and from country to country. Table 7.1 presents a comparative situation of waste generated in 2004 and 2015 for European Union member countries (kg per capita).

From the data presented, in 2015 compared to 2004, there is an increase in generated waste (kg per capita) mainly for the following countries: Croatia (+29.28%), Latvia (+27.04%), Slovakia (+26.05%), and Lithuania (+20.11%). Also, for the same period under review, the amount of waste generated per capita fell particularly in the following countries: Bulgaria (−30.05%), Romania (−29.23%), Spain (−27.67%), and Ireland (−20.35%). However, in 2015, the countries with the high-

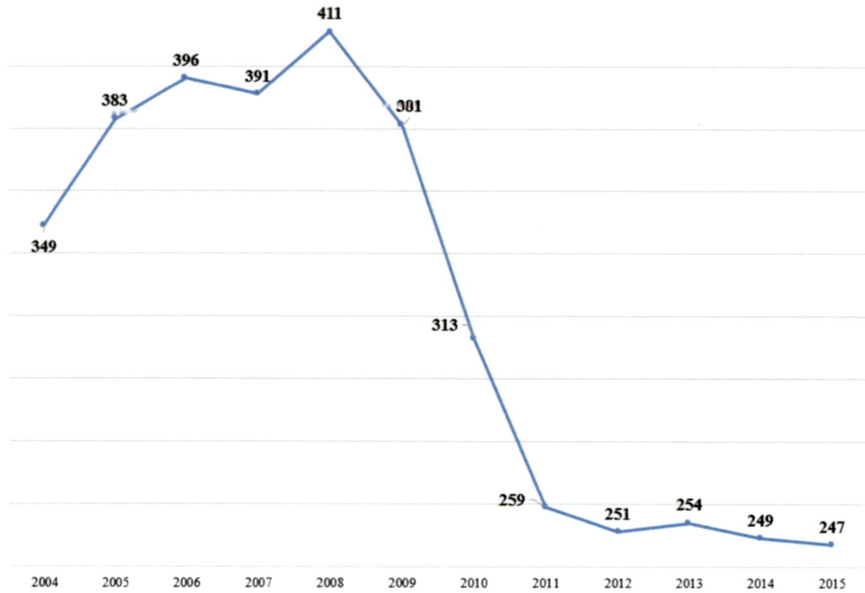

Fig. 7.1 The evolution of waste quantity generated per capita (*Source*: Data compilation done by the authors by using EUROSTAT)

est recorded waste per capita are Denmark, Cyprus, Germany, Luxembourg, and Malta. All of these countries have over 600 kg of waste generated per capita. Romania and Poland register the smallest waste generated per capita in the year 2015 – below 300 kg. At the same time, less than 400 kg of waste generated per capita is observed for the following countries: Czech Republic, Slovakia, Estonia, Hungary, and Croatia. In Romania, for the period 2004–2015, the evolution of the amount of waste generated per capita is shown in Fig. 7.1 (kg per capita). It is noted that for the period 2004–2008, the amount of waste generated per capita has increased almost continuously to 411 kg per capita. Also, for the period 2008–2015, there is a decrease in the amount of waste generated per capita, up to the value of 247 kg per capita.

Figure 7.2 shows the evolution of the quantity of waste treated per capita in Romania for the period 2006–2015 (kg per capita).

From the data presented, it is noted that the period 2007–2008 and the period 2012–2014, were periods when the amount of waste treated per capita increased. Also, the period 2006–2007 and the period 2008–2012 were periods when the amount of waste treated per capita decreased. The trend of recent years is a slight decrease in the amount of waste treated per capita. Taking into account the data presented in Fig. 7.1, it is noted that for Romania in the year 2015, 87.45% of the amount of waste generated per capita is treated. Furthermore, in this research paper, we will approach what is needed to be done to solve the issue of storing waste by recycling or incinerating it.

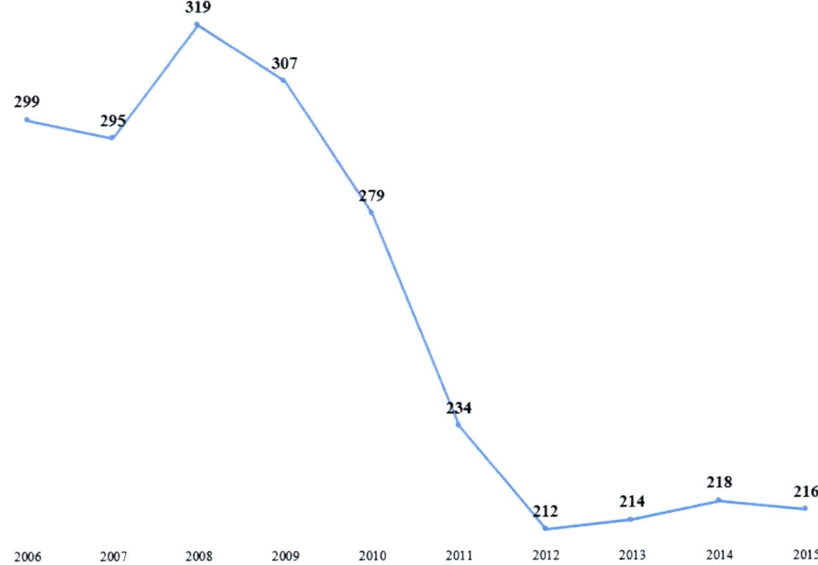

Fig. 7.2 The evolution of waste quantity per capita that was treated (*Source*: Data compilation done by the authors by using EUROSTAT)

7.5 Recycling and Incineration of Waste

7.5.1 Waste Recycling

Humanity has always been concerned with the collection, processing, and use of degraded products by the first consumers. Recycling determines and depicts all existing recycling means (paper, cardboard, plastics, glass, metals, and wood). The focus of waste recycling is on the efficiency of waste management policies through recovery but also through recycling, referring to the relationship between the quantities of recycled waste and the amount of waste delegated for recycling operations. The first organized units for recovering the reusable resources date back to Romania in 1949. With the help of the local public administration, large quantities of recyclable materials have been recovered, which has led to a lower ecological imbalance. The development of recycling of materials used in the Romanian economy starts from the following concepts: the awareness of both the population and the economic agents regarding the importance of materials recycling is low but can be improved by means of publicity, technological and technical support is relatively low at national level, and the value of the reusable materials is increased.

Recovery of plastics is a barrier to sorting by sorting, sorting according to specific mass and degraded technologies impeding the achievement of superior product quality. Plastic is a product made from gasoline, coal, and oil, and most of the materials used to make the plastic result from the oil refinery, so it is the problem that we use the resources we otherwise dispose of. Paper recycling can be done four times at most and brings a natural gain because it reduces the cost of industrial water, air pollution, electricity, water pollution, and the reduction of huge timber.

7.5.2 *Waste Incineration*

Composting is a well-known and long-established process and resolves a waste selection to take only those wastes that contain large amounts of organic matter, making them a bit simple: spraying and reaming on concrete platforms in a range of 3–4 weeks. Composting can be set in use, being the most environmentally friendly process of reentering waste in nature, as it does not cause noxiousness during processing and is used without modification in agriculture and horticulture (Rojanschi and Bran 2002, p. 194). When fixing and building the composting plant, it is necessary to prevent the propagation of flies and rodents and the spread of unpleasant smell and dust and to prevent the pollution of soil, air, and water. In order to prevent them from occurring, the capacity of the plant must be large, so that the residues are processed continuously. A certain distance of protection must be ensured between localities and the placement of the composting plant (Bold and Mărăcineanu 2006, p. 292). Waste incineration is the best way to neutralize garbage; from the sanitary point of view, it is considered to be the only acceptable hygienic process and the best economic solution is the burning of uncollectible waste (Negreaet al. 2007). The advantages of incineration of waste are that combustion plants can be located near the localities, reducing transportation costs and burning them at high temperatures, leading to a much faster and more complete neutralization, the pollution of the environment being small. The disadvantage is the noise generated by the heavy traffic of the transport vehicles and the high investment and operating costs. Also, a problem is the storage of residues resulting from the combustion process.

7.6 Conclusion

Changes that have taken place at the global level have made urgent action both internationally and nationally to ensure sustainable development, maintaining ecological balance. Each week, a citizen of the European Union produces about 5 kg of waste, an average of 36.7 metric tonnes of waste per year is managed on an annual basis, and 40% is recyclable material from the total municipal waste, about 20% of the uncontaminated waste is recovered. When it comes to waste management, we think about the complex problems that are happening at country level and that need an initiative of coordinated actions at the regional and local level, cooperation between civil societies and authorities and cooperation between states and with representatives of governments. In recent years, with the country's economic change and policy, consumption has also changed, so there has been a restructuring of industry and trade, with a much wider range of offers, consumer goods lasting shorter life, and more expensive packaging. An increase in commodity consumption has been identified due to the fact that the way of life has improved, and the behavior of the purchasing population has changed, as well as increasing the quantities of waste. As a result of the debate on waste-related issues, an increase in the level of involvement and problem-solving in waste management was noted in urban agglomerations in 2016. However, efforts by authorities to expand the selective collection system covering sanitation services in order to improve the waste transport parks and increasing the level of absorption from European funds addressed to the integrated waste management section. New solutions must also be found to educate and raise awareness among the population, to increase the degree of collaboration between the actors involved in the urban waste management process. We believe that municipalities should be set up, a sector dedicated to waste management issues, so that European and national environmental funds can be accessed. For the new EU Member States (Romania, Bulgaria, and Croatia), the coverage of sanitation services has set the following progress: in urban areas, 55% in 2015 and 58.07% in 2016 and in rural areas, 38.6% in 2015 and 54.9% in 2016, so that efforts should be stepped up to establish and operate sanitation services in order to reach the local level targets (European Council 2017). Another problem is the lack of progress in reducing the amount of landfilled waste, but the composting of biodegradable waste in rural areas should be supported. In conclusion, much greater emphasis should be placed on waste avoidance procedures and measures to encourage the prevention of waste generation in the pursuit of general waste management concepts. Avoiding the formation of waste can lead to a reduction of the environmental effects.

Worldwide there is a sharp increase in the population. An effect of this increase is also the increase in the amount of waste. Good waste management can also be achieved by identifying new natural resources that support the current lifestyle.

Better management of the environment can also be based on the use of more efficient production processes. Thus, operating costs may decrease, but the degree of dependence on raw materials can also be reduced.

Partnerships and the promotion of good practice become necessary measures in order to properly manage the sensitive issues created by waste generation.

References

Bold OV, Mărăcineanu G (2004) Depozitarea, tratarea și reciclarea deșeurilor și materialelor (translation: Depositing, treating and recycling waste and materials). Matrix Rom, Bucharest

Bold OV, Mărăcineanu G (2006) Depozitarea, tratarea și reciclarea deșeurilor și materialelor (translation: Depositing, treating and recycling waste and materials). Matrix Rom, Bucharest

Ionescu S, Lefter V (2013) Perfecționarea managementului organizațiilor din România în domeniul valorificării deșeurilor regenerabile în condițiile integrării în Uniunea Europeana (translation: Improving organizational management in Romania for the domain of capitalizing renewable waste in the the parameters of entering the European Union). ASE Publishing, Bucharest

National Institute for Research and Development (Institutul Național de Cercetare-Dezvoltare pentru Protecția Mediului – I.C.I.M. Bucharest (2005) Metode și tehnologii de gestionare a deșeurilor, Colectarea și transportul deșeurilor și a materialelor reciclabile (translation: Methods and technologies of managing waste. Collecting and transporting waste and recyclable materials). http://www.deseuri-online.ro/new/download/Colectaretransport.pdf
Negrea A, Cocheci L, Pode R (2007) Managementul integrat al deşeurilor solide orăşeneşti (translation: Integrated management of urban solid waste). Politehnica Publishing, Timişoara
Rojanschi V, Bran F (2002) Politici și strategii de mediu (translation: Environmental policies and strategies). Economica Publishing House, Bucharest
Rojanschi V, Bran F, Diaconu G (2002) Protecția și ingineria mediului (translation: Environmental protection and engineering). Economica Publishing House, Bucharest
Rojanschi V, Bran F, Grigore F (2004) Elemente de economia și managementul mediului (translation: Elements of economics and environmental management). Economica Publishing House, Bucharest
***National Agency for Environmental Protection. http://www.anpm.ro/
***The European Committee for Economy and Social – Empowerment for the Directive of the European Parliament and Council for promoting using energy from renewable sources, Brussels, 28.07.2017
***The European Council – Recommendation for the Council's Recommendation for the National Program for Reform of Romania for year 2017 that includes the directive of the Council regarding the Convergence Program of Romania for year 2017, Brussels, 15.06.2017
***EUROSTAT database. http://ec.europa.eu/eurostat

Chapter 8
Municipal Solid Waste Management Practices in the Central Part of Libya

Abdelnaser Omran and Abdelsalam O. Gebril

Abstract A solid waste management (SWM) system includes the generation of waste, storage, collection, transportation, processing, and final disposal. It is a basic public necessity, and this service is provided by respective local bodies (LBs) in Libya. This chapter aims to examine the current status of municipal solid waste management practices in Ajdabiya city located in the central part of Libya. In fact, all Libyan cities are still struggling to achieve the elements of all MSW such as generation, collection, disposal, recycling, etc. Survey data are gathered from a random sample of Ajdabiya city. Descriptive statistical analyses are conducted using the Statistical Packages for Social Sciences (SPSS) software program. Findings have shown that majority of the residents were dissatisfied with the current existing solid waste management program in the city. This study also found that the current SWM services are inefficient. It was also found that improper solid waste management caused environmental pollution and significantly impacts the quality of life of nations as well as deteriorated public health in these cities.

Keywords Solid waste · Collection · Transporting · Disposal · Public health · Environment · Ajdabiya city · Libya

8.1 Introduction

Solid waste management has been a major challenge in towns and cities throughout the world and most especially in developing countries (Jin et al. 2006). The World Health Organization (2000) defined waste as something which the owner no longer

A. Omran (✉)
School of Economics, Finance & Banking, Universiti Utara Malaysia,
Sintok, Kedah, Malaysia
e-mail: naser_elamroni@yahoo.co.uk

A. O. Gebril
Department of Natural Resources and Environment, Faculty of Arts and Science, EL-Marj Branch,
University of Benghazi, EL-Marj, Libya

A. Omran, O. Schwarz-Herion (eds.), *The Impact of Climate Change on Our Life*, https://doi.org/10.1007/978-981-10-7748-7_8

wants at a given time and which has no perceived market value. In recent years, there have been several studies conducted on the influential factors affecting solid waste management system and programs in cities of developing countries. These studies were aimed at addressing solid waste management problems related to generation, collection, handling, incineration, dumping, landfilling, and recycling activities. However, most of these studies came out with useful solutions and recommendations. In Libya, wastes are being generated mainly from residential, commercial, and institutional land uses (Omran et al. 2011; Gebril et al. 2010). A solid waste management (SWM) system includes the generation of waste, storage, collection, transportation, processing, and final disposal. It is a basic public necessity, and this service is provided by respective local bodies (LBs) in Libya. This study attempts to examine the present status of waste management and its effects on public health and the environment in Ajdabiya city. In Libya, most of the cities if not all are still struggling to achieve the elements of all MSW such as generation, collection, disposal, recycling, etc. Mismanagement of municipal solid waste has caused many negative impacts to the environmental and is also a public health risk in the country. Improper solid waste collection and handling, inadequate landfill disposal, and the absence of recycling practices are some of the main factors that contribute to infrastructural challenges. In Ghana, Asase et al. (2009) noted that there was a lack of proper disposal sites in the country. Unprotected and uncontrolled dumps, which pose a danger to the public health, environmental health, and waste renewable resources and jeopardize residential development in these areas, are a commonality found in many developing countries. Ayomoh et al. (2008) had listed few problems related to improper landfill operation including health deterioration, accidents, flood occurrences, pollution of surface and underground waters, unpleasant odor, pest infestation, and gas explosion. Unlike developed nations, third world countries lack sanitary landfills, and often disposal sites are located at a considerable distance from communities. Rowland and Wenner (1994) asserted that inadequate collection and disposal of solid waste is a major factors contributing to the spread of gastrointestinal and parasitic diseases, which are caused by the proliferation of insects and rodents.

Anjum and Deshazo (1996) reported that in most cities, municipalities, and towns in developing countries, SWM costs consume between 20% and 50% of municipal revenues. However, the waste collection service levels remain low with only between 50% and 70% of the residents receiving services and most of the disposal being unsafe. Bhatia and Gurnani (1996) had further observed that the efficiency of collection of waste in urban areas of developing countries varies from 59% to 82% suggesting that a substantial amount of solid waste remains uncollected. Poor solid waste collection and disposal is a threat to public health and reduces the quality of life for urban residents. Several approaches have been suggested to improve SWM in developing countries. Research conducted by Gebril et al. (2010) in Benghazi city (Libya) concluded that population growth, rapid urbanization, and industrialization resulted in increasing the problem of solid wastes in Benghazi city. MSW management becomes a leading subject and is one of the main public concerns, since an inadequate waste management may result in sanitary and environmental problems (Magrinho et al. 2006; Şchiopu et al. 2007; Damghani et al. 2008; Altawati 2013). In a study conducted by Onibokun and Kumuyi (1999) in the city of Ibadan in Nigeria, the city is found to be polluted with decaying solid waste which is found everywhere in the city including the street drains and water bodies due to unavailability of SW facilities and lack of maintenance.

In Bangladesh, Hai and Ali (2005) studied the solid waste management system of Dhaka City Corporation (DCC) and found that DCC was not able to offer the desired level of services with the existing capacity and trend of waste management. Zurbrigg and Ahmed (1999) stated that problem areas in developing countries include (i) inadequate service coverage and operational inefficiencies of services, (ii) limited utilization of recycling activities, and (ii) insufficient landfill disposal. Sharholy et al. (2008) presented a comprehensive review of the characteristics, generation, collection and transportation, and disposal and treatment technologies of municipal solid waste practiced in India. The study is concluded with a few suggestions for the efficient management of such waste such as involvement of public and private sectors through NGOs, increasing the public awareness, proper timing and scheduling for collection of waste from house to house, proper design and placing of collection bins, and proper maintenance of transport vehicles for such wastes. Omran et al. (2017) conducted in the city of Al Bayda, located in the eastern part of Libya, as there have been several major problems facing the city in dealing with solid waste management. One of these problems is the generation, collection, handling, transportation, recycling, and disposal of municipal solid waste. They identified that lack of resources and services that significantly affects the disposal of waste, an inadequate number of waste collection containers making the distance to these containers for many households excessive, was the main leading factor to an increasing likelihood of dumping solid waste in open areas and roadsides.

Vidanaarachchi et al. (2006) conducted a study in Sri Lanka on the issues and challenges of solid waste faced in the country's Southern Province, and they found that only 24% of the households had regular access to the waste collection and that in rural areas it was less than 2%. They also found that a substantial number of households in areas without waste collection expected local authorities to collect their wastes.

A lot of literature has documented how an adequate legal framework contributes positively to the development of the integrated waste management system (Omran et al. 2007, 2011, 2017; Asase et al. 2009; Hazra and Goel 2009; Altawati 2013), while the absence of satisfactory policies (Mrayyan and Hamdi 2006) and weak regulations (Seng et al. 2010; Omran et al. 2009, 2017) are detrimental to it. About the pricing for disposal, Scheinberg (2011) noted that there are indications that high rates of recovery are associated with tipping fees at the disposal site. High disposal pricing has the effect of more recovery of waste generated that goes to the value chains or beneficial reuse of waste. Also, Pokhrel and Viraraghavan (2005) mentioned that insufficient financial resources limiting the safe disposal of waste in well-equipped and engineered landfills and absence of legislation are also factors contributed to the pricing for disposal. Also, Chung and Poon (2001) agreed that having a clear structure of charges for waste collection and disposal in place may even work as an incentive for waste reduction. They believe that there is a need to change the approach for waste reduction from the "command-and-control" to the use of economic incentives and "polluter pays" (Chung and Poon 2001). Another supportive by Parrot et al. (2009) discussed the statutory, financial, and physical aspects of MSW management in the city of Yaoundé, the capital of Cameroon. They identified transportation distances, infrastructure quality, and accessibility as decisive factors on waste collection considerations in this city.

Concerning recycling, Omran et al. (2009) investigated attitudes of households toward recycling solid waste in the state of Kedah in Northern Malaysia. They conclude that simple improvements at bring sites could significantly increase recycling rates. If dwellers and shopkeepers are given waste storage containers of a standard size and collection is done regularly, then people are less likely to throw waste onto roadsides (Omran et al. 2009). In Thailand, Suttibak and Nitivattananon (2008) investigated the factors that influence the performance of solid waste recycling programs, and their results reveal that there are some common significant factors affecting SWM and solid waste recycling, including the perception of administrator awareness of SWM problems and source separation. Regarding school garbage banks, the provision of monetary incentives, including interest and compensatory goods for recycling members, transportation costs, and low investment costs, significantly affects recycling performance. In another study, González-Torre and Adenso-Díaz (2005) reported that social influences and altruistic and regulatory factors are some of the reasons why certain communities develop strong recycling habits. The authors also showed that people who frequently go to the bins to dispose of general refuse are more likely to recycle some product at home, and in most cases, as the distance to the recycling bins decreases, the number of fractions that citizens separate and collect at home increases. Minghua et al. (2009) stated that to increase recycling rates, the government should encourage markets for recycled materials and increasing professionalism in recycling companies.

Factors like poor and inefficient coverage and operation of services, inadequate or missing recycling strategies and activities, and limited or unproductive management of wastes were mentioned by some scholars as factors contributing to poor solid waste systems in any country (Henry et al. 2006; Vidanaarachchi et al. 2006; Omran and Gavrilescu 2008; Longo and Wagner 2011). Other factors mentioned by other researchers are financial support for recycling projects and infrastructures (Nissim et al. 2005; Moghadam et al. 2009), recycling companies in the country (Omran et al. 2011, 2017; Tai et al. 2011; Henry et al. 2006; Altawati 2013), drop-off and buyback centers (Matete and Trois 2008) and organization of the informal sector (Sharholy et al. 2008; Guerrero et al. 2013), inadequate or missing recycling strategies and activities (Omran et al. 2011; Vidanaarachchi et al. 2006; Omran and Gavrilescu 2008), and limited or unproductive management of wastes (Omran et al. 2007; Omran and Gavrilescu 2008; Altawati 2013). From the above-reviewed studies, it can be summarized that solid waste management is affected by many different factors. Libya is facing a sharp contrast between its increasingly urban population and available services and resources. Solid waste management (SWM) is one such service where Libya has an enormous gap to fill. The current existing SWM services such as collection, transportation, disposal system, landfilling, and recycling in the city are inefficient. Such improper services for SWM did contribute many negative things in the city like deteriorating the public health, causing environmental pollution, and causing climate change which positively impact the quality of life of residents. Based on the above preceding facts, this study is conducted to ascertain the factors that will sustain the municipal solid waste management and practices in Ajdabiya city, Libya.

8.2 Description of the Study Area

Ajdabiya is a city situated in central northern Libya near the Mediterranean Sea coast at the eastern end of the Gulf of Sidra (Fig. 8.1). It is located on an arid plain about 6.4 km (4.0 mi) from the sea and is approximately 850 km (530 mi) from the Libyan capital of Tripoli and 150 km (93 mi) from Libya's second largest city, Benghazi. The city is the site of an important crossroad between the coastal road from Tripoli to Benghazi and inland routes south to the oasis at Jalu and east to Tobruk and the border with Egypt. Ajdabiya lies close to the Sabkhat Ghuzayyil, a large dry region below sea level. The town is divided into three parts, namely, North Ajdabiya, West Ajdabiya, and East Ajdabiya.

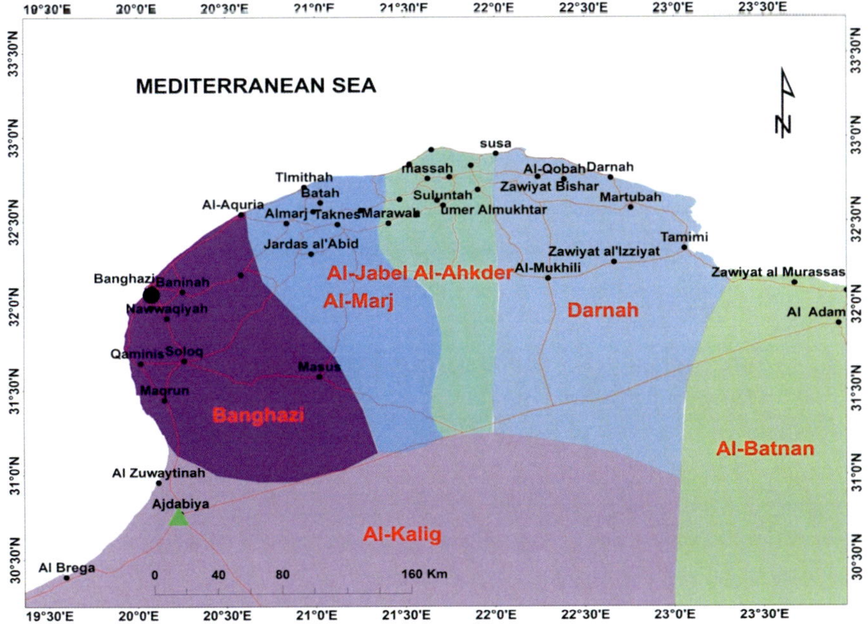

Fig. 8.1 Shows the location of Ajdabiya city in green color

8.3 Research Method

A descriptive research design with survey method was applied to carry out this study. A questionnaire was adopted from some conducted studies by Omran (2008), Omran et al. (2009), and Mahmoud (2013); the main objective of the survey is to identify the factors that could sustain the municipal solid waste management and practices in Ajdabiya city, Libya. Prior to distribution of the questionnaire, pretests were conducted in order to ensure item clarity and non-ambiguity. The targeted respondents are the residents in the abovementioned city. Out of 300 sets of a questionnaire distributed, 204 sets of the questionnaires have been fully done and returned which were useable for the findings analysis. Thus, the return ratio was 68%, representing a highly satisfactory level of participation when compared to similar postal surveys (Hansmann et al. 2006). All data were analyzed using the Statistical Package for the Social Sciences (SPSS), Version 20.0 for Windows software. The alpha level was set at 0.05 to determine statistical significance, and a descriptive statistics with a number of tests were used including chi-square tests.

8.4 Results and Analysis

8.4.1 Background of the Respondents

Table 8.1 shows the characteristics of the respondent base on their demographic background. Almost all of the respondents were Libyan. They were from various groups of age; the highest came from those aged between 26 and 35 years old, with 39.7%, followed by 27% from those aged between 26 and 35 years old. Almost 63% (62.7%) of them were males, and 37.3% were females. The majority of the respondents have finished the university level (30.9%). 45.1% of the respondents were working with the public sector, and nearly 30% were unemployed (students).

8.4.2 Satisfaction with the Existing Municipal Solid Waste Management Services

Respondents were asked if they were ever satisfied with the current SWM services in their areas. The chi-square analysis reveals that satisfaction level of SWM services is significantly related to age, education level, and occupation (p-value < 0.05) as shown in Table 8.2.

Respondents were asked regarding their knowledge and awareness about the effects of unconcerned and indiscriminate solid waste disposal on current issues in the chosen city (Table 8.3). They stated their awareness of these effects on their public health (30.9%) and on environmental pollution (47.1%).

Table 8.1 Background of the respondents

	Percentage (%)
Age	
10–15 years old	2.9
16–25 years old	39.7
26–35 years old	27
36–49 years old	23
50–69 years old	6.9
70 years and above	0.5
Gender	
Male	62.7
Female	37.3
Educational level	
Preliminary school	3.4
Primary school	7.8
Secondary school	21
Medium Diploma	15.7
High Diploma	14.2
BSc degree	30.9
MSc degree	4
PhD degree	–
Did not attend any school	3
Occupation	
Public sector	45.1
Private sector	5.4
Own Business	5.9
Student	29.9
Unemployed	13.7

Table 8.2 Respondents' satisfaction with solid waste management services based on demographic factors

	Chi-square results	
Demographic	χ^2	p-value
Gender	13.887	0.083
Age	35.385	<0.018**
Educational level	16.025	<0.042*
Occupation	22.342	<0.034*

**Statistically significant at $p < 0.05$*
***Statistically significant at $p < 0.01$*

About the respondents' knowledge on the solid waste collected by the designated body in the cities in Libya (Table 8.4), most of the respondents (49%) agreed that SW is treated and disposed of in landfills, while 36.3% of the participants indicated that SW is incinerated in the open. Also, there were some respondents who were aware that solid waste is also dumped into the sea after treatment (11.8%).

Table 8.3 Knowledge of the effects of unconcerned solid waste disposal

Factors	%
Effect of unconcerned solid waste disposal on your personal health	13.7
Effect of unconcerned solid waste disposal on public health	30.9
Effect of unconcerned solid waste disposal the economy	8.3
Effect of unconcerned solid waste disposal on environmental pollution	47.1

Table 8.4 Ways to dispose of solid wastes by the designated body in the city

Factors	Percentage (%)
Solid waste is treated and disposed of in landfills	49
Solid waste is disposed of without treatment in landfills	6.9
Solid waste is dumped at sea after treatment	11.8
Solid waste is dumped at sea without treatment	3.9
Solid waste is incinerated in controlled facility	21.6
Solid waste is incinerated in the open	36.3
No idea	13.2

Table 8.5 illustrates the respondents' preference concerning the groups who should perform SW collection and disposal. Among the respondents, 42.6% agreed that SW collection and disposal should be performed by the private sector. However, 37.7% preferred the public sector (government) and 33.8% preferred nongovernment organizations (NGOs).

The study found the factors that hinder the respondents from participating in SWM to keep the city clean. As presented in Table 8.6, it can be found from the results that the most important factor was the irregular waste collection (mean, 4.02); improper disposal of waste collected by the authorities (mean, 4.0) was another factor. Third on the list was the lack of transparency in waste management (mean, 3.47).

8.4.3 Specific Problems Related to Solid Waste in the City

The participants were questioned to whether there are any specific problems related to solid waste in the city, and the question was offered with two choices, either yes or no. However, it was seen that with the majority of the respondents, 82.4% had declared that there are many specific problems related to solid wastes in this city, while the other 17.6% said "no." However, when those stated that they had a particular problem, asked on classifying these types of problems, 32.8% of them

Table 8.5 Preference groups to handle solid waste collection and disposal

Factors	Percentage (%)
Public sector should handle solid waste collection and disposal	37.7
Private sector should handle solid waste collection and disposal	42.6
Nongovernment organizations (NGOs) should handle solid waste collection and disposal	33.8

Table 8.6 Factors that hinder the respondents from participating in SWM

Factors	Mean	SD	Ranking
Irregular waste collection	4.02	1.35	1
Lack of transparency in waste management	3.47	1.19	3
Improper disposal of waste collected by waste-collecting authorities	4.00	1.15	2

Table 8.7 Some specific problems of solid wastes in the city

Items	Percentage (%)
Accumulation	32.8
The lack of containers	35.8
Distort the general appearance of the waste	31.4
Others	–
Total	100

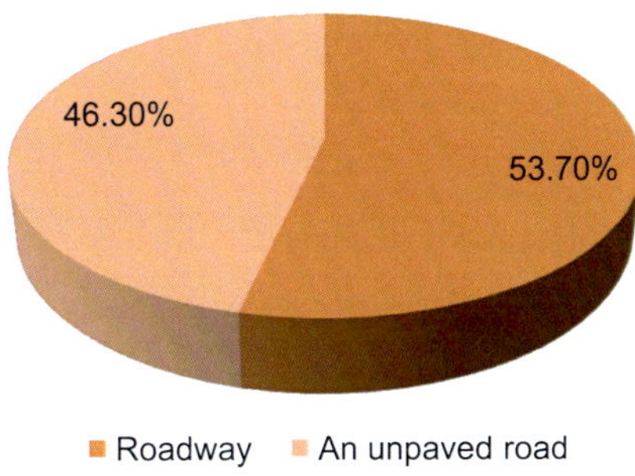

Fig. 8.2 Access roads to the residential areas in the city

mentioned that accumulation of the waste is one of the biggest problems, followed by lack of containers (35.8%) (see Table 8.7).

Concerning the road leading to the house, about 58.7% of the respondents indicated that they do not have an easy connection or suitable roads to their residential areas, while other respondents (41.3 %) did report that they had road but in an unpaved condition (Fig. 8.2).

Table 8.8 The distance of the road from the location of the houses

Options	Percentage (%)
<4 m	17
4–6 m	18.4
6–8 m	18.4
>8	46.2
Total	100

Table 8.9 The numbers of containers near to house

Quantity	Percentage (%)
1	14.2
2	9.9
>3	6.9
No containers	69
Total	100

8.4.4 Determining the Distance of the Road from the Residential Areas

Residents were asked to estimate whether the road is located closer to the residential area or not. As a result, the analysis of the findings showed that 46.2% mentioned that it is more than 8 m from their houses. It was also noticed that 18.4% of them stated that the distance is ranged from 4 to 6 m and 6 to 8 m, respectively. Other findings are given in Table 8.8.

8.4.5 Quantities of the Containers and Bins in the City

The local authorities in the city of Ajdabiya placed bins and containers in certain areas. The participants were asked on the number of containers located to their houses. However, it was found that 14.2% of the residents indicated that the number of containers is only one and nearly 10% mentioned that their streets have only two containers (see Table 8.9).

8.4.6 Issues Related to the Distance of the Containers from the Residential Areas

Concerning the distance, the participants were questioned about the currently available distances for the containers from their houses, and they had been asked to estimate the current distance by giving them some estimated choices including

Table 8.10 The currently estimated distance for container that you are ready to cut them to deliver the waste to the container

Options	Percentage (%)
10–20 m	10.7
20–50 m	8.3
50–100 m	3.4
100–150 m	3.5
More than 150 m	7.4
No container	66.7
Total	100

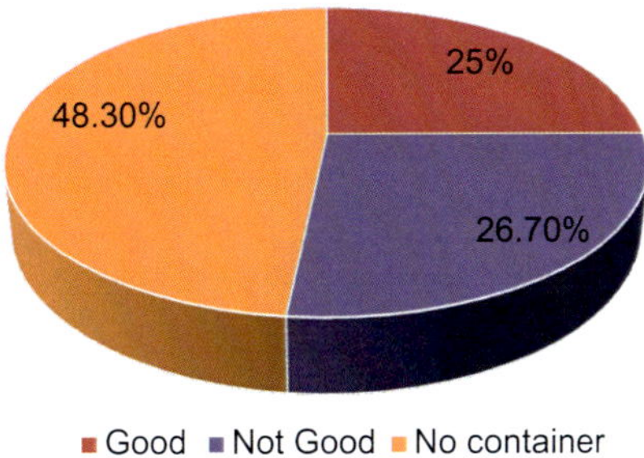

Fig. 8.3 Locating the container regarding suitable places

10–20 m, 20–50 m, 50–100 m, 100–150 m, and more than 150 m. Therefore, as a result, it is found that 66.7% of the respondents are complaining that there was no container provided in their residential area. Other respondents (10.7%) had indicated that these containers were located in the distance ranged 10–20 m (Table 8.10). This can be attributed to the fact that these containers are located in an unarranged manner and varied from residential area to another.

8.4.7 *Location of the Containers*

The participants in this survey were requested to give their perceptions and opinions on the design and shape of the container regarding environmental location. The survey found that 48.3% of the respondents proved that there were no containers in their area and about 26.7% stated they were not placed in a good way (see Fig. 8.3).

In addition to its environmental location, the residents were also questioned on the design and shape of these containers in terms of its health aspects. The results showed that 25.5% of the respondents were unhappy with these containers regarding its health aspect (see Fig. 8.4).

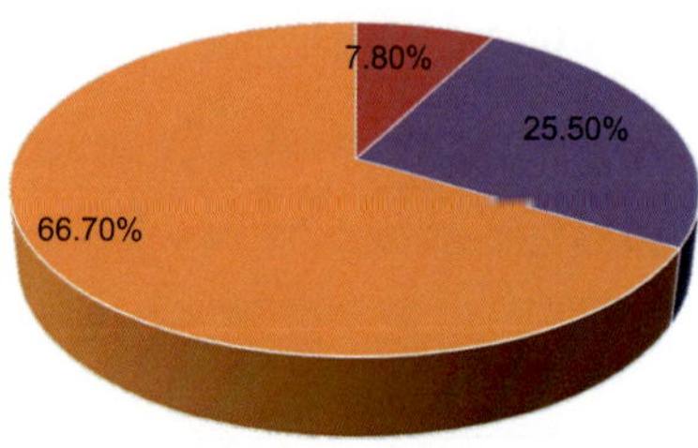

Fig. 8.4 Form the container regarding health

Table 8.11 The number of times for unloading the container

Periods	Percentage (%)
Daily	8.8
Day after day	10.3
Every three days	6.4
Once a week	5.8
Takes more than a week	2
No container	66.7
Total	100

Table 8.12 Knowledge on finding the containers with a full of waste

Items	Percentage (%)
Yes	68.2
No	13.2
Sometimes	18.6
Total	100

8.4.8 Period of Unloading of the Containers

The respondents were asked about the period of unloading container in the city of Ajdabiya. In this respect, about 66.7% of them indicated that there were no containers at all in their areas, and only 10.3% mentioned that the unloading process is day after day, while 8.8% reported the unloading process of these containers is daily (refer to Table 8.11).

The respondents were asked whether the containers are always found with a full of waste or not. As shown in Table 8.12, the analysis found that quite a high percentage of the participants (68.2%) declared that these containers are always full of waste and 18.6% of them answered with sometimes these containers were found to be full of waste.

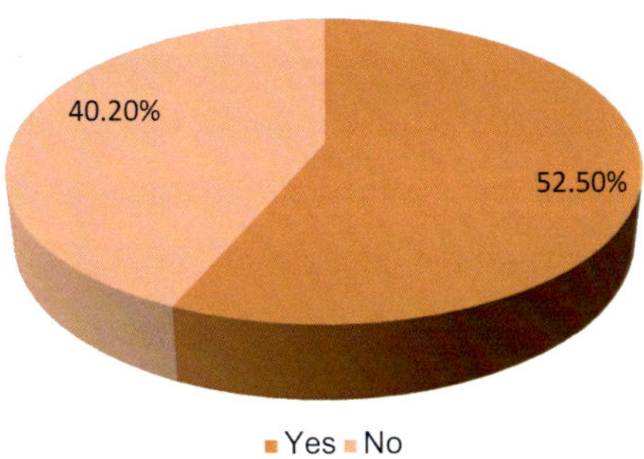

Fig. 8.5 Spray insecticide into the containers

Table 8.13 Willingness fees to be paid for improving solid waste collection services

Costs	Percentage (%)
2–3 LYD	13.2
4–6 LYD	27.5
6–8 LYD	–
More than 8 LYD	59.3

8.4.9 Spraying and Insecticide Inside the Containers

The respondents were asked about their ideas regarding the government activities and responsibilities in spraying any insecticide into the containers or not. However, a vast percentage of the respondents (40.2%) complained that these activities are not practiced (Fig. 8.5). As concerned, this protection is usually practiced in the summer season where it used to avoid the spread of the breeding of insects.

8.4.10 Willingness to Pay Costs for Services

During the surveying, the households were asked about how much fees they can afford to pay to improve the services of solid waste collection. However, the analysis of the survey showed that about 59.3% of the respondents would be able to afford to pay more than 8 LYD, followed by 27.5% of the respondents who agreed on paying 4–6 LYD. These findings are clearly presented in Table 8.13.

Table 8.14 Respondents' awareness of SWM laws and regulations

Items	Frequency
Awareness of law/regulation	
Yes	20.1
No	79.9
Support the law/regulation	
Yes	93.6
No	6.4
Preference punishment	
Community work	15.2
Fine	75.5
Imprisonment	9.3
Others	–

Table 8.15 Respondents' awareness of tax and contribution for SWM

Items	Percentage
Awareness on tax	
Yes	22.1
No	77.9
Support the tax	
Yes	80.9
No	19.1

8.4.11 Awareness of Laws and Regulations Implemented Concerning SW

In this section, the respondents were asked about their knowledge and awareness of any existing laws and regulations on solid waste management in their living area in the three cities. However, the results indicated that nearly 80% (79.9%) of them were unaware of the laws and regulations that promote discipline among people who pollute their cities. The survey had further asked the respondents whether they agreed to support such regulations or laws to be implemented or not! As shown in Table 8.14, it can be seen that 93.6% of them supported the laws and regulations prescribing appropriate penalty for people who pollute the city where the results indicated that 75.2% of the participants agreed that the appropriate punishments to those who pollute the three cities are to be paid a fine and only 9.3% of them suggested the imprisonment as one of the appropriate punishments.

As shown in Table 8.15, the respondents indicated their awareness on tax payments for the upkeep of the mentioned cities. The majority (77.9%) was unaware of the tax, but almost all respondents (80.9%) agreed by saying that paying the tax can keep the cities clean.

8.5 Conclusion

Municipal solid waste management remains a major environmental concern in the Libyan cities in general and in the city of Ajdabiya in particular. This study concluded that the mentioned city is facing deficiencies in varying degrees, and these problems include inadequate coverage, lack of short- and long-term environmental plans, lack of both human and financial resources, and improper choice of technology and planning. Therefore, there is an urgent need to improvise the situation of MSW management practices to stop further decay and deterioration in this city. These improvements can be achieved if the following recommendations are taken into serious consideration by the responsible body for sustaining solid waste management in the mentioned city. These are the following:

1. The current existing regulations for SWM are not properly enforced, and SWM seems to be considered of low priority. Therefore, possible laws and policies should be reinforced, and offenders must be considered.
2. Illegal dumpsites and the much-uncollected solid wastes are rising throughout the city. Such unacceptable practices will pose many serious health problems to the residents. Therefore, the authorities must find ways to reduce these unexpected risings.
3. Lack of community participation in SWM across the city was commonly noticed. Therefore, involving the community in SWM activities will have a positive input to the city.
4. Raising public awareness in the city needs to be created especially at the generator level to minimize solid waste generation.
5. Specific policies and regulations to sustainable solid waste management in the city addressing all types of waste and with the clarity of the roles and responsibility of each citizen should be developed.
6. Sanctions and penalties of solid waste mismanagement should be put in place and strictly followed.
7. Involving the private sector involvement, as well as other actors in SWM, needs to be increased as this will definitely play a big role in improving the efficiency of SWM.
8. Monitoring system for SWM monitoring should be put in place to ensure adherence to SWM regulations/laws.

References

Altawati MA (2013) Developing a sustainable solid waste framework in Al-Bayda City, Libya Unpublished PhD thesis. Universiti Sains Malaysia, Penang, Malaysia

Anjum M, Deshazo J (1996) Household demand for improved solid waste management: a case study of Gujranwala Pakistan. Water Dev 24(5):857–868

Asase M, Yanful EK, Mensah M, Stanford J, Amponsah S (2009) Comparison of municipal SWMS in Canada and Ghana: a case study of the cities of London, Ontario, and Kumasi, Ghana. Waste Manage 29(10):2779–2786
Ayomoh MKO, Oke SA, Adedeji WO, Owaba OEC (2008) An approach to tackling environmental and health impacts of municipal solid waste disposal in developing countries. J Environ Manage 88(1):108–114
Bhatia MS, Gurnani PT (1996) Urban waste management privatization. Reaching the unreached: challenge for the 21st century. Twenty-second WEDC conference. Discussion paper
Chung S-S, Poon C-S (2001) Accounting for the shortage of solid waste disposal facilities in Southern China. Environ Conserv 28:99–103
Damghani AM, Savarypour G, Zand E, Deihimfard R (2008) Municipal solid waste management in Tehran: current practices, opportunities, and challenges. Waste Manage 28(5):929–934
Gebril AO, Omran A, Pakir AK, Aziz HA (2010) Municipal solid waste management in Benghazi (Libya): current practices and challenges. Environ Eng Manage J 9(9):1289–1296
González-Torre B, Adenso-Díaz PL (2005) Influence of distance on the motivation and frequency of household recycling. Int J Waste Manage 5:15–23
Guerrero LA, Maas G, Hogland W (2013) Solid waste management challenges for cities in developing countries. Waste Manage 33:220–232
Hai F, Ali M (2005) A study on solid waste management system of Dhaka City Corporation: effect of composting and landfill location. UAP Journal of Civil and. Environ Eng 1(1):18–26
Hansmann R, Bernasconi P, Smieszek T, Loukopoulos P, Scholz RW (2006) Justifications and self-organization as determinants of recycling behaviour: the case of used batteries. Resour Conserv Recycl 47:133–159
Hazra T, Goel S (2009) Solid waste management in Kolkata, India: practices and challenges. Waste Manag 29(1):470–478
Henry RK, Yongsheng Z, Jun D (2006) Municipal solid waste management challenges in developing countries – Kenyan case study. Waste Manage 26(1):92–100
Jin J, Wang Z, Ran S (2006) Solid waste management in Macao: practices and challenges. J Waste Manage 26:1045–1051
Longo C, Wagner J (2011) Bridging legal and economic perspectives on interstate municipal solid waste disposal in the US. Waste Manage 31(1):147–153
Magrinho A, Didelet F, Semiao V (2006) Municipal solid waste disposal in Portugal. Waste Manage 26(12):1477–1489
Mahmoud A (2013) Developing a framework for sustainable municipal solid waste management in Libya: a case study in Al-Bayda city. PhD Thesis, Universiti Sains Malaysia, Malaysia
Matete N, Trois C (2008) Towards zero waste in emerging countries – a South African experience. J Waste Manage 28:1480–1492
Minghua Z, Xiumin F, Rovetta A, Qichang H, Vicentini F, Bingkai L et al (2009) Municipal solid waste management in Pudong New Area, China. J Waste Manage 29:1227–1233
Moghadam AM, Mokhtarani N, Mokhtarani B (2009) Municipal solid waste management in Rasht City, Iran. Waste Manage 29(1):485–489
Mrayyan B, Hamdi MR (2006) Management approaches to integrated solid waste in industrialized zones in Jordan: a case of Zarqa City. J Waste Manage 26:195–205
Nissim I, Shohat T, Inbar Y (2005) From dumping to sanitary landfills solid waste management in Israel. J Waste Manage 25:323–327
Omran A (2008) The attitude of Malaysian on recycling of solid wastes in Malaysia case studies in the major towns of the East Coast and North Malaysia. PhD Unpublished thesis. Universiti Sains Malaysia, Pulau Pinang, Malaysia
Omran A, Gavrilescu M (2008) Perspective on municipal solid waste in Vietnam. Environ Eng Manag J 7(4):59–67
Omran A, Mahmood A, Aziz HA (2007) Current practice of solid waste management in Malaysia and its disposal. Environ Eng Manage J 6:295–300

Omran A, Mahmood M, Aziz AH, Robinson GM (2009) Investigating households' attitude towards recycling of solid waste in Malaysia: a case study. Int J Environ Rese 3(2):275–288
Omran A, Alsadey S, Gavrilescu M (2011) Municipal solid waste management in Bani Walid, Libya: practices and challenges. Environ Eng Manag J 2(4):228–232
Omran A, Altawati M, Davis G (2017) Identifying municipal solid waste management opportunities in Al-Bayda City, Libya. Environ Dev Sustain. https://doi.org/10.1007/s10668-017-9955-3
Onibokun, Kumuyi (1999) Governance and waste management in Africa. In: Adepoju G, Onibokun (ed) Managing the monster urban. Southeastern Nigeria. Theoretical and Empirical Researches in Urban Management, 3(12), pp 137–148
Parrot I, Sotamenou J, Dia BK (2009) Municipal solid waste management in Africa: strategies and livelihoods in Yaoundé, Cameroon. Waste Manage 29(2):986–995
Pokhrel D, Viraraghavan T (2005) Municipal solid waste management in Nepal: practices and challenges. J Waste Manage 25:555–562
Rowland AJ, Wenner P (1994) Environmental health. Edward Arnold, London
Scheinberg, A (2011) Value added: modes of sustainable recycling in the modernisation of waste management systems. PhD Wageningen University, Netherlands
Şchiopu A, Apostol I, Hodoreanu M, Gavrilescu M (2007) Solid waste in Romania: management, treatment, and pollution prevention practices. Environ Eng Manage J 6(5):451–465
Seng B, Kaneko H, Hirayama K, Katayama-Hirayama K (2010) Municipal solid waste management in Phnom Penh, capital city of Cambodia. Waste Manage Res 29:491–500
Sharholy M, Ahmad K, Mahmood G, Trivedi RC (2008) Municipal solid waste management in Indian cities – a review. Waste Manage 28:459–467
Suttibak S, Nitivattananon V (2008) Assessment of factors influencing the performance of solid waste recycling programs. Resour Conserv Recycl 53(1/2):45–56
Tai J, Zhang W, Che Y, Feng D (2011) Municipal solid waste source-separated collection in China: a comparative analysis. J Waste Manage 31:1673–1682
Toulmin S (1953) The philosophy of science: an introduction. Hutchinson University Library, London
Vidanaarachchi CK, Yuen TS, Pilapitiya S (2006) Municipal solid waste management in the Southern Province of Sri Lanka: problems, issues, and challenges. J Waste Manage 26:920–930
WHO (2000) Health and environment in sustainable development five years after the earth summit WHO/EHG/97.8
Zurbrig C, Ahmed R (1999) Enhancing community motivation and participation in solid waste management. Published by SANDEC News no.9

Chapter 9
Fixing Climate Change: Accounting Disclosure Remedies

Ahmad Bello

Abstract Many scholars attributed industrialization and rapid urbanization to economic development; however, they failed to consider the price of the said development that could be attributed to the negative impact on the environment via carbon emission. Emissions are documented to have devastating effect on climate condition. The carbon emission could also be a result of corporate entities' role on industrialization and urbanization where their actions are seen as the contributing factor to global warming, hence, climate change. Thus, the issue of Kyoto Protocol arose to curtail carbon emission. While not all the countries agreed to the Kyoto agreement on reducing carbon emissions among others, enforcing sustainability reporting among firms across the globe could have a positive impact toward saving the earth. This piece explores areas in which sustainability reporting could be expanded beyond corporate environmental reporting to other necessary disclosures that will curve recklessness in the course of pursuing economic goal.

Keywords Climate change · Corporate disclosures · Sustainability reporting

9.1 Introduction

Climate change is a global term attributed to increase or decrease in climate due to the emission of carbon 4 oxide which destroys ozone layers, hence the change in climate (Akpomuvie 2011). While carbon is a scientific word that could be carbon dioxide, carbon 4 oxide, and carbon mono oxide, among other elements, the process of releasing the gas to atmosphere is called emission (Lal 2004). The emission of carbon is one of the fundamental and key global problems that attracts global attention on recent years which needed immediate attention in order to address the issue of global warming as stated ab initio. The said emission could be from flaring of gas and releasing of carbon as a result of firm production which result in disposing of waste gases through the use of combustion (Boxall et al. 2005). In the same vein,

A. Bello (✉)
Department of Accounting, School of Business, Ahmadu Bello University, Zaria, Nigeria

A. Omran, O. Schwarz-Herion (eds.), *The Impact of Climate Change on Our Life*, https://doi.org/10.1007/978-981-10-7748-7_9

climate change could be as a result of associated gas wastages which emulate from production of crude oil by oil and gas firms; thus, industrialization and urbanization could also be causes of carbon emission which in turn result to global warming. The said emission is a crucial problem worldwide, and it is expressively drawing attention internationally, particularly on the matter of global warming. So many factors could lead to increase in carbon that has tremendous effect on our society today. Researchers assert that industrialization and the activities of firms are two of the major contributors that led to global warming resulting from carbon emission. The emission could also be as a result of crude oil production (Akpomuvie 2011). While the world is facing the demons born by the firms today, many of the firms are not accountable for their role in contaminating the environment through their activities. However, the world is so much concerned that lead to the Kyoto Protocol with the aim of curtailing carbon emission (Akpomuvie 2011). Yet, government and firms pay less attention and even ignore many recommendations of the Kyoto Protocol, and this could be attributed to the laxity of the global standard concerning environment, which includes sustainability reporting. While other countries pay attention to the sustainability reporting of their firms, however, many countries including Nigeria are not worried in terms of the reporting aspect of sustainability of their firms.

Sustainability reporting is necessary as it is an indication of commitment on economic, social, and environmental issues concerning firm and their related environment. Whereas many advanced economies consider it as mandatory yet; other countries like Nigeria consider it not even optional but render it a waste of time since there is no regulatory framework or even ordinary reporting concerning firms in relation to either social or environmental issues that are related to sustainability of the firms. Despite the sluggish attitude by the regulatory bodies on firms reporting and the voluntary nature of the sustainability reporting by the Nigerian firms, however, some but limited firms do have partial sustainability reporting. It is partial in the sense that not all components of sustainability are considered in their reporting.

To meet up the persistent increase in demand for a standard reporting in relation to sustainability issues, firms operating in the highly industrialized nations increase their reporting standard relating to economic, social, and environmental issues which are all embedded in sustainability reporting of firms as prescribed by the Global Reporting Initiative (GRI) guideline. However, Nigeria should not be in isolation even though the country is not recognized as an industrialized nation yet; many firms in Nigeria do contribute negatively to the environment as a result of their operation among which oil and gas companies are said to be environmentally sensitive than other firms as prescribed by their operational behavior in which the production of the oil results to an emission of associated gas called gas flaring (Orubu et al. 2004).

9.2 The Relevance of Climate Change Study

Industrialization and rapid urbanization couldn't have been achieved without a price of negatively destroying environment, which in turn has a devastating effect on climate. According to Ismaila et al. (2016), climate change is the key factor influencing environmental disclosure by companies. The effect of economic activities has implication on the environment; companies especially in the oil and gas industry are prone to activities impacting negatively on the environment.

Corporate entities are major agents of industrialization and urbanization. The duos are responsible for climate change. While not all the countries agreed to the Kyoto agreement of reducing carbon emissions among others, enforcing sustainability reporting among firms across the globe could have a positive impact toward saving the earth we live. This piece explores areas in which sustainability reporting could be expanded beyond corporate environmental reporting to other necessary disclosures that will curve recklessness in the course of pursuing economic goal.

9.3 Epistemology, Materials, and Methods

The research idea was formulated and guided by three interconnected theories that explained environmental degradation, multinational firms' activities, and general firms' behaviors to environmental reporting and regulatory accounting. Thus, the pollution haven hypothesis (PHH) predicts that, under free trade, multinational firms will relocate the production of their pollution-intensive goods to developing countries, taking advantage of the low environment monitoring in these countries (Umed 2006). Over time, developing countries will develop a comparative advantage in pollution-intensive industries and become "havens" for the world's polluting industries. Secondly, legitimacy theory, which seeks to relate firms' behavior voluntarily to exact honor and respect as it grows and develops (Chan et al. 2014). While on the other hand, in mandatory regulatory accounting theory, Dye (1990) posits that in an unregulated information market, firm being the sole supplier of their financial report can choose to underproduce the information (both in content and context) for it to charge exorbitant price. This undersupply of information would not be in the public and economic interest, as such mandatory reporting becomes necessary. Regulation to an extent of exacting full disclosure of environmental issues might force firms to sustainability disclosure. This chapter is purely conceptual and analytical. Analysis of normative documents, systematization and generalization of facts and concepts, and more importantly synthesis and analysis of literature with a view to deduce and extract facts for meaningful inference formed the basic frame of the paper. Where necessary some pictorial evidences as well as numerical fact were giving to buttress and support the deductive analysis.

9.4 Literature Analysis: Facts and Results

The main thrust of the study was to analyze and synthesize literature evidence on the extent and drivers of climate change and seek a regulatory accounting solution to containing the menace. Initially the study pre-hypothesizes that quest for rapid industrialization, neglect to piece of environmental degradation, and political cultures are attributable key drivers to climate change. Furthermore, it is believed that these drivers can be redress with accounting regulations, increasing the requirements of corporate disclosure and stock market incentive. This section dwells in analysis and synthesis of quantitative and qualitative facts to prove the hypothesis.

9.4.1 Carbon Emission

Among the area where this study is interested to be incorporated in the sustainability reporting is the amount of carbon emission released to the atmosphere by the companies as a result of their operation (Fig. 9.1). This should be mandatory as the action is totally against humanity where few are benefiting from the said emission of gas which, of course, causes global warming among others, while reporting nothing in relation to carbon but only concentrated on the other parts of sustainability reporting which is economic and the social aspect with few environmental portion. Thus, there is a need for the companies to report any carbon release by their firm be it directly or indirectly.

Fossils fuel would continue to dominate the power sources in Nigeria unless there is an adequate measure to invest in renewables and other clean energy alternatives. Coal and generator emissions account significantly to impact on climate (Fig. 9.2).

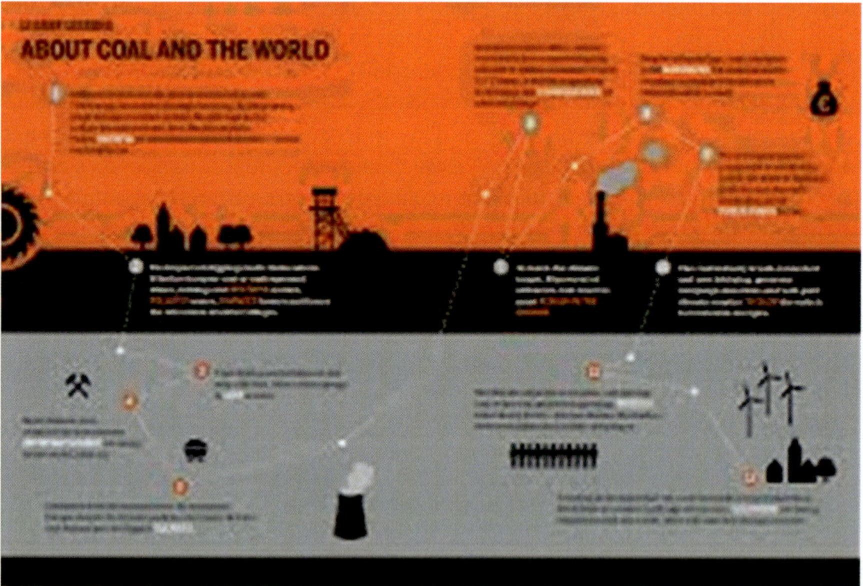

Fig. 9.1 Coal emission (Source: WTO 2015)

The World Resources Institute estimates that Nigeria's greenhouse gas emissions in 2012 (its latest record for the country) exceeded 296 MtCO2e (excluding land use). The figure exceeds 474 MtCO2e when land use is included (CAIT Climate Data Explorer 2015). The data available for the period 2000–2012 show an upward trend in emissions with drops in levels for 2007 through 2009, compared to 2006 values (280 MtCO2e without land use and 463 MtCO2e with land use). However, the emissions in 2011 and 2012 were each 16 points (excluding land use) and 11 points (including land use) higher than 2006 levels.

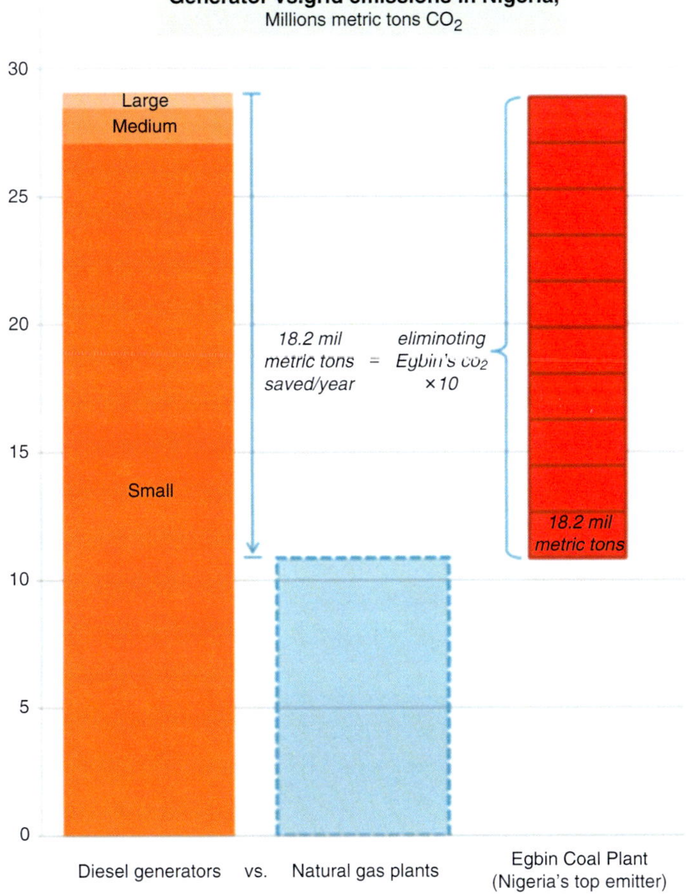

Fig. 9.2 Generator vs carbon emission (Source: Todd and Madeliene 2014)

9.4.2 Flaring of Gas

Another important aspect of reporting standard especially by oil-producing firms is gas flaring. Gas flaring is a common practice by oil-producing firms in Nigeria even though that has been addressed globally. The said gas flaring can be handled in three either ways: (a) to harness for use as liquefied natural gas, (b) to reinject into the earth, and finally (c) to vent or flare it (Bassey 2008). The latter is not economically, socially, and environmentally friendly, yet many companies find it easy to flare than utilizing it as a result of the high cost attributed to utilization of associated gas (Malumfashi 2007). Despite the role played by the firms on polluting the environment, yet the firms did not take the reporting aspect of the gas flaring as important since they pay penalty and the reporting is not mandatory but rather voluntary. David and Victor (2017) document that emission of carbon dioxide CO_2 is statistically significant and positively related to manufacturing capacity utilization of Nigeria both in the short run and long run. Figure 9.3 shows how gas flaring can be handled.

To add more information on the role played by many firms in Nigeria in terms of flaring of gas, however, its found that as at 2004, oil firms flare and vent gas equivalent to the combined gas consumption of Central and South America annually (World Bank Group 2004). This is also attributed to Africa as it also flares gas equivalent to half its power consumption (Anomohanran 2012). In addition, Nigeria is among the highest countries that flare gas globally (Ishisone 2004). Moreover, in the report of the World Bank, it was reported that natural gas is flared annually to the tune of 25% of gas consumption in the United States and 30% of European Union

Fig. 9.3 Gas flaring in Niger Delta (Source: Olukoya 2016)

gas consumption equivalent. Meanwhile, 13% of the flares of the said gas in the world come from Nigeria as prescribed by the World Bank. With all this quantity that is more than enough to meet Nigeria's energy needs, yet many firms in Nigeria could not report their role in terms of the quantity of flared gas released to the atmosphere in their financial reporting (Okoji 2000). Through this obnoxious act, the country could lose $2.5 billion annually if the maintenance of the environment, health of the public, and even the agricultural backwardness that is a result of climate change are estimated (Malumfashi 2007).

9.4.3 *Deforestation*

As opposed to afforestation, deforestation is advertent or inadvertent action to reduce the amount of trees in the forest (Fig. 9.4). It is being obseved that in the savannah and Guinean savannah of the Northern Nigeria, desert is approaching by five kilometers yearly (IOA 2015). This is attributed to bush burning and excessive cutting of trees for firewood and other industrial usages. Over 1 million trees are cut in Northern Nigeria without replacement, and combined with bush burning, this is destroying the environment. Desert encroachment is rapidly moving toward the savanna belt region by approximately 5 kilometers yearly.

Fig. 9.4 Deforestation (Source: www.Vanguardngr.com 2017)

9.4.4 Challenges

Continuing dependency on fossil fuels and wood as energy sources, paucity of climate change policies and programs, and changing peoples' behavior are the key identifiable emission reduction challenges. It is worthy to note that Nigeria was not a signatory at thet Paris Agreement Intended Nationally Determined Contributions (INDCs). This was hitherto the general view of many developing economies such as China that believe most international pacts to curve emission are retrospective and might dwarf their economic growth.

9.4.5 Remedies

9.4.5.1 Regulatory Accounting and Standardization

The framework of regulatory accounting which stresses on the role of corporate financial reporting plays in stimulating corporate executive behavior and effective means of enforcing certain desired direction on firms. Increasing concern of environmental and social impacts of organizational activities has raised the need for disclosure on environmental and social issues (Lamberton 2005; Gutherie and Farneti 2008). Furthermore, dissatisfaction of the activities of multinationals has increased the need for additional nonfinancial disclosure and metrics. While in developed economies' mandatory disclosure on environmental and social issues is a measure of concern, emerging economy activities of multinationals are not fully checked fixating on economic value in addition to the detriment of sustaining the environment. Nigeria has the highest record of environmental degradation due to activities of multinational oil firms in Niger Delta. Soil is compromised for oil. Water, fauna and flora are being polluted. Despite the records the firms operate under limited liability enterprises limiting creating a shadow of non-compliance with public limited companies' provisions and requirements. Environmental disclosure is very poor among firms in Nigeria as documented by many studies; information relating to social and governance issues dominates the pattern of reporting in cosmetic manner for firms to laud their image and avoid problem mostly with host communities (Johnson-Rokosu and Olanrewaju 2016). They further provide empirical evidence on firms studied providing little quantifiable data to improve their corporate image and to be seen as responsible corporate citizens. This is consistent with the study of that sustainability reporting disclosure practice in Nigeria is still very ad hoc, general, and self-laudatory in nature.

Regulatory bodies such as Securities and Exchange Commission of Nigeria have many weaknesses among which implementation of their standard on all firms is usually very poor. Even though there is standard in relation to general reporting in Nigeria, there is limited standard concerning reporting of carbon emission and climate change. As documented by so many studies, Nigeria is among the highest gas flares in the world (Orubu et al. 2004), yet it is increasingly recognized by a wide range of stakeholders, including the petroleum industry itself, which flaring and venting of gas constitute a waste of economically valuable resources and very dangerous to the environment which could be curtailed (Nwokeji 2007: World Bank Group 2004). For that reason alone, a question can be raised as to the extent of how regulatory bodies consider the emission of gas or flaring of gas in Nigeria. Do the regulatory bodies also take into account in considering the price of the environmental damages caused by the operation of firms in Nigeria? That needs an answer.

Despite the record of operation of firms especially the oil companies that have vast association and non-associated gas reserves, the production to meet up the demand of the consumers in Nigeria is low. The commercial expoloration of gas in relation to oil is low, leading to persistent carbon emission. (Aghalino 2009; Bankole 2001). It was also found out by the World Bank that developing countries account for more than 85% of gas flaring and venting, with Nigeria, Iraq, and the Islamic Republic of Iran as the highest flares (Malumfashi 2007). Thus, there is a need for more proactive measures amongst which the standard of reporting should be addressed to incorporate all the environmental necessary information that could improve the environmental issue, hence curtailment of climate change. This can only be done if the regulatory bodies set some standards and make it mandatory for firms operating in Nigeria to implement the said standards in respect of the sustainability reporting.

9.4.5.2 Coercive Measures

Despite the economics of alternative energy sources than fossils, Nigeria and many emerging economies continue using high-emitted sources. According to opinion by ICF (2016), generator power plant and the use of small generators for households and corporate electric power have more devastating effects to the environment in terms of CO_2 emissions (Table 9.1). Furthermore, Todd and Madeliene (2014) reported that with an estimated 9 million units of generators in use, Nigerian generators emit about 29 million metric tons of CO_2 each year. They further opined that by replacing generators with electricity from large-scale natural gas plants, Nigeria could cut its CO_{-2-} output by 18 million metric tons per year at a 63% cut. Moreover, the absence of political will to enforce environmental laws in the country has given leeway to economic agents. Environmental laws are established to mitigate the threatening environmental problems that emanate from human activities in the quest for economic growth and development (Hakeem and Joseph 2014).

To mitigate environmental degradation, policies ranging from ban of use of small generators to investment in gas turbine can curb the devastating effect. In addition, punitive measures on reckless cutting of trees by individuals and corporate organization need to be enforced to save the climate. Likewise, with trillion reserves in liquefied natural gas, enforcing the policy of "from flaring to energy" will save the country from huge emission.

Table 9.1 Comparison of CO_2 emission between generators and power plants

	Diesel generators	*CO_2 emissions g/kWh*
1	Small (<60 kW)	1580
2	Medium (>60 kW,<300 kW)	883
3	Large (>300 kW)	699
	Power plants	
1	Coal	969
2	Oil	792
3	Natural gas	553

Source: Todd and Medeliene (2014)

9.4.5.3 International Treaties and Agreements

Climate change aggravated by global warming has no place in geographical location. The effects are global irrespective of where it was emanated. Therefore, it's a moral and ethical responsibility to the whole world to join in the campaign as well as enforcement of treaties and agreements on cleaner environment. For advanced economies multinational firms that are seeking for safe haven, polluted countries should be monitored, and young poor nations should require an international treaty on pollution protection. Initiatives such as energy equity and energy per capita need to be legitimized internationally.

9.5 Conclusion

Environmental accounting and regulations would continue to be a topical issue, owing to continuing expansion in environmental problems, economics, communal disputes, and advancement in technology. The world we live in today with the passage of time would never be like before. Insatiable quest for input materials, intense competition, and developmental race will simultaneously affect environment and lead to climate change.

This chapter discusses the role of firms on environmental degradation as well as their standard of reporting in relation to the sustainability reporting. Despite the role played by the firms on Nigerian environment, which is negative especially the carbon emission, gas flaring which in turn destroys ozone layers, hence global warming, the mode of their reporting is not worthy for consideration with respect to sustainability reporting. Their actions could be attributed to the voluntary nature of the sustainability reporting as well as the lack of standard that could address the sustainability issues in Nigeria. Other factors could be political will of the government to address the ugly nature of the firm's activities toward environment even though this has been discussed in the Kyoto Protocol. Thus, this study recommends for Nigerian firms to pay more attention to their sustainability reporting which is to make sure carbon emission and gas flaring (where necessary) are reported accordingly. The study also deemed it necessary to recommend for setting of high standard rule concerning environmental issues as well as climate change issues and the enforcement of the said standard on Nigerian firms. Thus there is a need for Nigerian regulatory bodies on firms to formulate and implement highly competitive standard, which could be internationally recognized in line with the Kyoto Protocol on environmental issues and their respective operating firms. This action will help the government to control the emission of carbon, and that will save the huge amount of money the government uses to spend to control environmental hazard and degradation among others.

The paper recommends further as a policy input, developing and emerging economies need to design implementable policies for climate change, and the policy must be tailored toward mandatory financial disclosures on environmental degradation using sustainability reporting indices. Questions as to what extent does firm profit contribute to the environment need to be disclosed. Heavy regulations via banned and excessive tariffs on wooden materials should be encouraged. Laws and regulations from global and local stock markets should be set on a minimum reserve threshold based on firms' activity to be utilized as an investible fund for research and development in cleaning and sustaining environment. Lastly, climate change should be seen as global phenomena; it has no boundary to geographical location where the pollution emanates, as such international agreements and treaties should be enforceable and any country that creates a safe haven for pollutants should be isolated.

References

Aghalino S (2009) Gas flaring, environmental pollution and abatement measures in Nigeria. J Sustain Dev Afr 11(4):219–238

Akpomuvie OB (2011) Tragedy of commons: analysis of oil spillage, gas flaring and sustainable development of the Niger Delta of Nigeria. J Sustain Dev 4(2):200–210

Anomohanran O (2012) Determination of greenhouse gas emission resulting from gas flaring activities in Nigeria. Energy Policy 45:666–670

Bankole AS (2001) Incentives and regulations to abate gas flaring in Nigeria: towards an optimal gas utilization policy. In: Nigeria economic society (NES) Natural resources use, the environment and sustainable development, conference proceedings of Nigerian economic society pp 519–534

Bassey N (2008) Gas flaring: assaulting communities, jeopardizing the world. In: The environmental rights action in conjunction with Federal Ministry of Environment. Federal Mininstry of Environment, Nigeria

Boxall PC, Chan WH, McMillan ML (2005) The impact of oil and natural gas facilities on rural residential property values: a spatial hedonic analysis. Resour Energy Econ 27(3):248–269

CAIT Climate Data Explorer (2015) CAIT-historical emissions data (countries, U.S. States, UNFCCC). World Resources Institute, CAIT Climate Data Explorer, Washington, DC

Chan MC, Watson J, Woodliff D (2014) Corporate governance quality and CSR disclosure. J Bus Ethics 125(1):59–73

David IO, Victor ES (2017) The dynamics of carbon dioxide [CO_2] emission on Nigerian capacity utilization. Scholars Bulletin (A Multidisciplinary Journal) an official publication of "Scholars Middle East Publishers", Dubai, United Arab Emirates

Dye RA (1990) Mandatory versus voluntary disclosures: the cases of financial and real externalities. Account Rev 65:1–24

Gutherie J, Farneti F (2008) GRI sustainability report by Australian public sector organizations. Public Money Manag 28(6):361–366

Hakeem I, Joseph OT (2014) Rethinking environmental law enforcement in Nigeria. Beijing Law Rev 5:307–321

https://www.vanguardngr.com News (2017)

ICF International (2016) Nigeria summary in analysis of intended nationally determined contributions (INDCs) June 2016, USAID resources to advance LEDS implementation program, pp 35–36

Institute of Administration (2015) Poverty in northern Nigeria: causes and remedies, Research Studies. ABU, Zaria

Ishisone M (2004) Gas flaring in the Niger Delta: the potential benefits of its reduction on the local economy and environment. Retrieved on December. Retrieved from http://nature.berkeley.edu/classes/es196/projects/2004final/Ishone.pdf

Ismaila Y, Joseph S, Ekundayo O (2016) The extent of environmental Disclosures in listed Oil and Gas companies in Nigeria. Being a Proceedings of Nigeria's economic development conference titled "Nigeria's economic challenges since Independence: Issues and Prospects from Multidimensional Perspectives". Federal University Dutsinma, 6 to 8th June 2016. pp 400–413

Johnson-Rokosu SF, Olanrewaju RA (2016) Corporate sustainability reporting practice in an emerging market: a case of listed companies in Nigeria. Azerbaijan journal of knowledge horizons - economics and. Soc Stud 8(2):148–156

Lamberton G (2005) Sustainability accounting- a brief history and conceptual framework. Account Forum 29(1):7–26

Malumfashi GI (2007) Phase-out of gas flaring in Nigeria by 2008: the prospects of a multi-win project. Nigeria gas flaring petroleum. Train J 4(2):1–39

Nwokeji U (2007) The Nigerian national petroleum corporation and the development of Nigerian oil and gas industry: history, strategy and current directions. A reserch report by James A. Baker III Institute for Public Policy of Rice University. Available @ www.bakerinstitute.org

Okoji MA (2000) Petroleum oil and the Niger delta environment. Int J Environ Stud 57(6):713–723

Olukoya OAP (2016) Gas flaring, climate change and the Nigerian case. International Disaster and Risk Conference IDRC 2016

Orubu C, Odusola A, Ehwarieme W (2004) The Nigerian oil industry: environmental diseconomies, management strategies and the need for community involvement. J Hum Ecol 16(3):203–214

Rasmussen CE, Williams CKI (2006) Regression. Gaussian processes for machine learning, Chapter 2. https://doi.org/10.1093/bioinformatics/btq657

Todd M, Madeliene G (2014) How can Nigeria cut CO_2 emission by 63%? Build more power plants. Views from the Center of Global Development

Umed T (2006) Pollution haven hypothesis or factor endowment hypothesis: theory and empirical examination for the US and China. Working paper series, CERGE-EI. available @ www.CERGE-ei.cz

Uwuigbe U (2012) Web based corporate environmental reporting in Nigeria: a study of listed companies. Informatica. Economica 16(3):27–36

World Bank Group (2004) Global gas flaring reduction: A public-private partnership. A GGFR project brief, Available @ www.worldbank.org

World Trade Organization (2015) International trade statistics 2015. World Trade Organization. https://www.wto.org/english/res_e/statis_e/its2015_e/its2015_e.pdf

Chapter 10
Impacts of Climate Change on Agriculture in Malaysia

Wen Chiat Lee and Amir Hussin Baharuddin

Abstract Climatic change is a reality that faced by the world that seems to be failed understanding of the phenomena. Human being can reduce the impacts of climatic change. They can do so if they realize the need to work together. The authors of this chapter attempt to show how human activities can increase the negative impacts of climatic change by way of lifestyle. For example, human beings contribute to surface water runoff and flooding. Human beings can also increase the carbon dioxide content of the air from fuel burning. The options would be alternatives or greener fuel and reducing the number of cars on the road. In terms of the alternatives, we must increase the solar research and applications and manage the nitrogen and carbon cycle through understanding of the environment and the ecosystem. The alternatives and options are exposed in the book chapter. Moreover, this chapter also tries to reduce the unnecessary fear of the climatic impacts on human beings by suggesting steps and alternatives and possible international cooperative efforts.

Keywords Climate change · Lifestyle · Agriculture · Malaysia

10.1 Introduction

Malaysia is a tropical country that is located in Asia with potentials for crop production. Malaysia is the second largest exporter of oil palm in the world and third largest rubber producer in the world (United States Department of Agriculture 2017). Fruit trees such as jackfruit, cempedak, breadfruit, bananas, durians, and coconuts are planted extensively in Malaysia. According to the World Bank, 24% of the total land in Malaysia is agriculture in year 2014. Malaysia has an area of 78,390 square km in year 2014, with 7.6 million hectares of arable land suitable for crops.

W. C. Lee · A. H. Baharuddin (✉)
School of Economics, Finance and Banking, Universiti Utara Malaysia,
Sintok, Kedah, Malaysia
e-mail: amirhussin@uum.edu.my

A. Omran, O. Schwarz-Herion (eds.), *The Impact of Climate Change on Our Life*, https://doi.org/10.1007/978-981-10-7748-7_10

Agriculture is very important to Malaysia because agriculture provides major food for Malaysians. However, Malaysia had not focused much on agricultural development since the implementation of the eighth Malaysian Plan in year 2001. Malaysia had become too dependent on food import. In year 2015, Malaysia imported a total of RM45 billions of food, which represents around 20% of the total budget of a country in that year (Carvalho et al. 2016). This means that Malaysia is not self-sufficient in food. Malaysia is turning their agricultural land into housing and buildings. Malaysia is fast urbanizing and declares 75% urban and 25% urban in 2016 (The World Bank 2016a). Fruit trees had been chopped off to make way for housing and urbanization in Malaysia. China had targeted 50% urbanization and 50% rural for the Chinese. China maintains their farmers at rural area, by giving lands to their farmers to plant crops to feed the urban people and for exports. China could not afford not to produce food and import food because the population in China is huge at 1.4 billion people. To make matter worse, Malaysia is more vulnerable to climate compared to China. Climate change can occur everywhere in the world. Climate change is a change in the typical or average weather of a region or city. This could be a change in a region's annual rainfall, a change in Earth's average temperature, or a change of typical precipitation pattern (NASA). Agriculture production sector is most directly impacted by climate change. Being a traditional agriculture country, Malaysian agriculture is thus vulnerable to the impact of climate change. Malaysia must assess the source of climate change and minimizing the impacts of climate change on agriculture to ensure sustainability of the agriculture.

10.2 Literature Review of Climate Change and Agriculture

This section reviews the relationship between climate change and agriculture. Climate change, agriculture production, and the contribution of the agriculture sector to the economy will be discussed. The success stories of agriculture in other countries such as Southeast Asia, Europe, and the United States are discussed as well to enrich the understanding of how these countries had prepared themselves for climate changes in the future with advances in technology, idea and adaptability of people, and crop and animal production.

10.2.1 Climate Change and Agriculture in Malaysia

Agriculture and climate are interdependent. Plants are the primary factories of agriculture. Plants take in carbon dioxide from the air through their leaves. They take in moisture and chemical substances from the soil through their roots. Out of these, plants make seeds, fruits, fibers, and oils that man can use by using the energy of the sunlight. Animals, on the other hand, depend on plants for their food; they can eat many parts of plants that man does not, such as stems and leaves of grasses. The

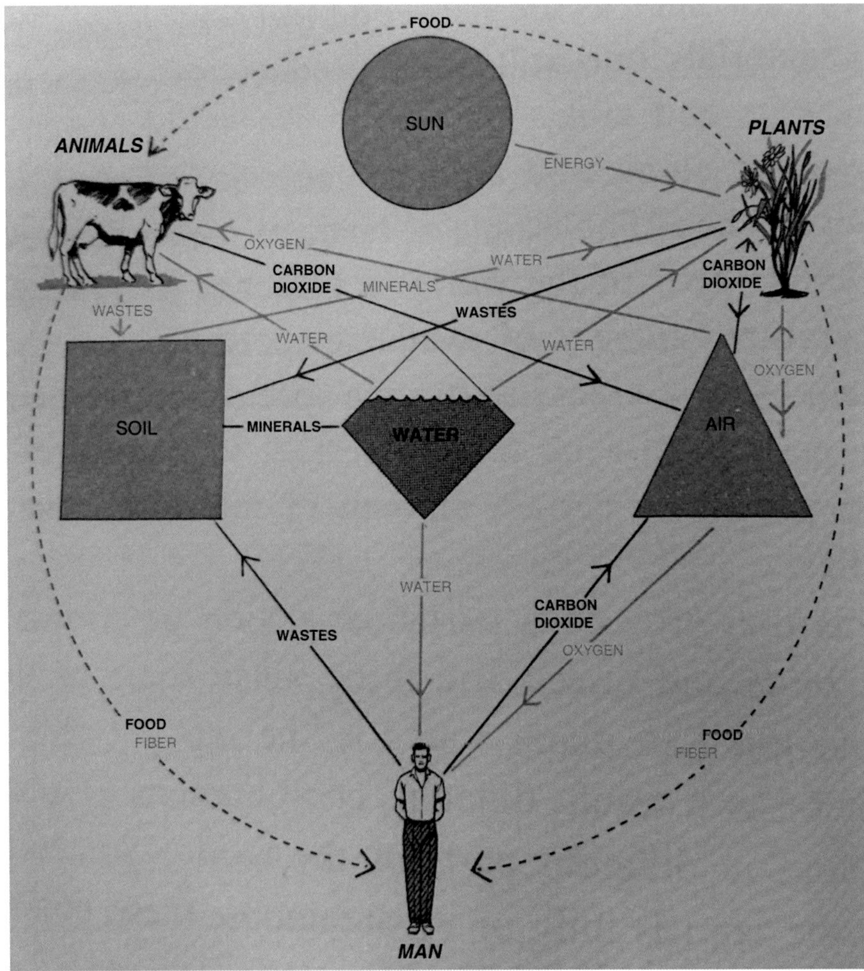

Fig. 10.1 Biological production process of agriculture (Source: Mosher 1966)

animals transform plant materials into other products of use to man: meat, hides, wool, eggs, and milk. The diagram of the nature of agriculture is shown in Fig. 10.1.

There are certain characteristics of the biological production process of agriculture that are beyond our control. Human beings do not create sun, water, air, and soil. The sun, water, soil, and air are the biological production factors that are needed by the plants and animals to grow. However, the growth of plants and animals is disrupted if human beings destroy the biological production factors such as the sun, water, air, and soil. Thus, human beings must understand the biological production process of agriculture and prevent from harming the biological factors through human activities such as rapid urbanization, accelerate house construction, chopping down trees, and flattening the mountains.

Climate changes can be nature-made and man-made. For nature-made climate change, Malaysia cannot do much about it and must be prepared to face the climate changes. For the man-made climate change, Malaysia can avoid it by changing production strategies and consumption styles. The man-made climate change is caused by urbanization, illegal loggings and development projects near the coastal areas and rivers, and careless exploitation of the resource and bad management practices especially land management.

Urbanization is the process by which towns and cities are formed. Increasing number of people begin living and working in the urban areas. When urbanization occurs, the cutting of forest trees and exploitation of land will accelerate. More buildings and houses are built to benefit mankind in the urban areas. Urbanization contributes to destruction of the forests and may contribute to man-made climate change. One of the significant examples in Malaysia is the conversion of paddy land to housing and urban centers and the destruction of major rivers.

In recent years, Malaysia has set its sight on the Vision 2020 Plan, which could propel the country to be a developed nation with high-income status at the gross national income of US$15,000 per capita. Many buildings, houses, and factories are built in order to speed up the Malaysian Vision 2020, to become developed nation. Thus, activities that harm the environment such as tree cutting from areas that were formerly forests for tree crop production and urbanization contribute to environmental damage. Land reclamation in Penang was executed for urban development. This caused damage to the ecosystem and agriculture especially when permanent structure was erected physically. The human quest for modern development has altered the local weather and climates, releasing greenhouse gases to the air and coal burning which release more toxic to the air. Human activities bring harm to the in situ climate, soil, and water degradation through conversion of paddy fields to build houses for the people have caused the agriculture sector to deteriorate. Table 10.1 shows that the Malaysian policies have shifted from agriculture in rural area to services and manufacturing which is located in urban areas.

The national focus on the agriculture sector of Malaysia has dropped as shown in the contribution of sectors to Malaysian GDP in year 1965 and 2015. The percentage of contribution of agriculture has dropped from 31.5% in year 1965 to only 8.9% in year 2015. Although the percentage of agriculture to GDP is only 8.9%, Malaysia must not abandon agriculture because agriculture represents the source of food to the people of Malaysia. Moreover, the agriculture sector also employed

Table 10.1 Contribution of sectors to Malaysian GDP in year 1965 and year 2015

Sectors in Malaysia	% to GDP in year 1965	% to GDP in year 2015
1. Services	43.9	53.5
2. Manufacturing	10.4	23
3. Agriculture	31.5	8.9
4. Construction	4.1	4.4
5. Mining	9.0	9

Source: Department of Statistics Malaysia (2016)

Table 10.2 Shows the contribution of agriculture sectors in year 2015

Subsector of agriculture	% to agriculture sector in year 2015
Oil palm	46.9
Livestock	10.7
Fisheries	10.7
Rubber	7.2
Forestry and logging	6.9

Source: Department of Statistics Malaysia (2016)

1.1 million people. The country that abandons agriculture has to import foods. Malaysia has imported about RM45 billion worth of foods in year 2015 because of not growing them. Malaysia has stagnant agriculture sector and focuses on exporting crops to other nations. Table 10.2 reflects that Malaysia does not focus on food production.

As shown in Table 10.2, it can be observed that the oil palm and rubber contribute around 55% to the agriculture production in year 2015. Malaysia planted 5.6 million hectares of oil palm (Malaysian Palm Oil Board 2016) and 1.07 million hectares of rubber (Malaysian Rubber Board 2016). Malaysia has a total of 7.6 million hectares of arable agricultural land. Oil palm plantation and rubber plantation have taken up 87.76% of the total agriculture land (6.67 million out of 7.6 million hectares). This means that Malaysia is not having enough farmland for other essential food crops. We cannot eat food from oil palm and rubber. Malaysia hopes to use money obtained from export of oil palm and rubber to import foods. But that is not enough as food import bill is growing very high. In the year 2015, it was at RM45 billion or 17% of total national budget expenditure at RM260 billion a year. We can learn from success stories of agriculture in Southeast Asia (Thailand and the Philippines), Europe (the Netherlands and Switzerland), and the United States of America (USA) if we want to contribute to local food sufficiency and global food security in the face of climatic changes.

10.2.2 Preparedness of Thailand and Italy

Thailand has successful agriculture sectors. Their agriculture is highly competitive, diversified, and specialized. Rice is the top crop in Thailand. Sixty percent of the 13 million farmers grow the crops (SCB Economic Intelligence Centre 2017). Agriculture contributes about 10% of the Thailand GDP (The World Bank 2016b). Forty percent of the labor force work in the agriculture sector. Agriculture provides important jobs to the Thai people. Thailand grows a lot of food crops such as durian, bananas, rice, lychee, and many fruits for local consumption and export (mostly to Malaysia). Thailand had not abandoned agriculture and become self-reliant in foods.

Italy has good planning on agriculture. The northern part of Italy has a strong tradition in pig and cattle breeding, whereas the southern part of Italy is allocated for wheat, tobacco, olives, stone fruit, sugar beets, and tomato cultivation. This shows that Italy has a good planning in agriculture. Industrial and developed country like Italy had not abandoned agriculture, even though about 74% of the Italian GDP is generated from their service sector (The World Bank 2016c).

10.2.3 Successful Agricultural Stories in the Netherlands and Switzerland

The Netherlands is a small country and only 1/8 the size of Malaysia. The climate is cold and not very suitable for crop production. It is surrounded by sea with very limited land. However, the Netherlands never abandon agriculture. The Netherlands has 2.5 million of cattle to produce milk and Malaysia only has 30,000 cattles to produce milk. The self-sufficiency of milk in Malaysia is 5% of local demand. Ninety-five percent of milk in Malaysia is imported from Australia, the Netherlands, and New Zealand. The Netherlands produces Dutch Lady Milk, and it has good agricultural universities in the world such as Wageningen University, Utrecht University, University of Twente, and University of Amsterdam that do extensive research on agriculture. Another country that never abandon agriculture is Switzerland. Switzerland mountains have cold climate. Most of the countries may not be suitable for tropical fruits. The geographical area of Switzerland is surrounded with alps and limited land for agriculture. But this does not undermine the Swiss to develop its agriculture. The Swiss cultivates herbal and produces herbal candy, named Ricola since 1930. The Swiss now sells Ricola to every part of the world. We can observe the vision and the planning of the country on developing agriculture and make the country great. This country may not be affected by massive climatic changes.

10.2.4 Success of Agriculture in the United States of America

The United States of America (USA) often asks the developing countries to develop industries and urbanize the people to be developed nation. However, the United States had never abandoned their agriculture because agriculture represents their source of food. Farm output only represents about 1% to GDP and employs 1.4% of the total US employment (United States Department of Agriculture 2016). The United States never abandons the agriculture and invests heavily on agriculture especially in science and technology, research and development, product development, and farmers' productivity. The United States has agricultural experimental stations in most states funded by the government and has the cooperative extension service to educate the farmers on crop plantation and changes of crop production and utilization strategies. The United States had always recognized the importance of technology to adapt to climatic changes.

10.3 Malaysian Development and Agriculture

In recent years, Malaysia had set her sight on Vision 2020 Plan, with the idea of propelling the country to be a developed and high-income nation with gross national income of US$15,000 per capita. Many buildings, houses, and factories were built in order to speed up Malaysia to become developed nation. The activities associated with the massive infrastructures built could harm the environment. Cutting of trees for urbanization and indiscriminate land reclamation such as those in Penang could bring damage to the ecosystem of the sea and fishery. The human quest for modern urban living has altered the local weather and climates. The advent of cars and the increase of lorries and buses had resulted in release of greenhouses gases to the air, and the act of coal burning released more toxic materials to the air. Human activities could do permanent harm to the climate, soil degradation, and the destruction of water supply.

Malaysia declares itself as 75% urban and 25% rural area to achieve high urbanization, industrial, and developed state in year 2020. Malaysia has emphasized on heavy industry, knowledge economy, and information communication technology (ICT) and services and to abandon agriculture. Agriculture happens in the rural area. If the rural area experiences area shrinkage, it is hard to plant crops and foods. Malaysia has to rely on imported food. Not all wrong policies on agriculture fall on mismanagement and wrong planning; young generation also plays a role in attributing to wrong policies on agriculture. Young people do not like to be farmers. They conceived agriculture as dirty, poor, and difficult. Young generation or generation Y would like to work in urban area and does not like to work as farm producers in rural areas. That is the reason why Malaysia has 80% urban and 20% population. The young generation does not like to assume the farming role of their parents. The older generation may need to sell the land to housing developers. The housing developers convert the agricultural land to housing land and build houses. This action alters the environment of agriculture and contributes negatively to climate change in Malaysia with prolonged dry seasons, intensive rains, and massive flooding of the country. The urban areas faced flash floods.

The government of Malaysia also does not build adequate facilities and infrastructure in small rural towns to encourage farmers to stay in rural area to plant crops. Facilities and infrastructure access in small town, agricultural processing facilities such as small and medium enterprises, and agricultural marketing facilities such as organized fresh market and kiosks to sell farm products are not there in small towns. Government had not given incentive or attention to retain people in agriculture. New towns are developed and old towns near to rural area are left declining with limited assistance. All the developments of infrastructure are focused on big cities only such as Kuala Lumpur, Johor Bahru, and Penang. Malaysian small rural towns do not have facilities like those found in agropolitan in Europe and America. In United Kingdom, the government gives incentives to the people to stay in rural area to produce food and prevent traffic and people congestions in urban area. The town planning is effective and the rural area is green. The climate change has little effects on the United Kingdom. The United Kingdom has large areas of green grass for the dairy cattle to consume and produce milk to feed the British. Their urbanization is focused and had not resulted in much man-made climatic change. Housing, urban infrastructure, and automobiles had not dramatically changed the living environments with ample food supply. They are prepared for climatic changes.

10.4 Impacts of Climate Change on Agriculture

United States Environmental Protection Agency (EPA) highlights that agriculture and fisheries depend very much on climate changes. Increases in temperature and carbon dioxide (CO_2) contents of the atmosphere can vary some crop yields in some places. Crop yields also depend on nutrient levels, soil moisture, and water availability. Droughts and floods can cause the crop yield to drop and food supply to shrink.

Climate change can cause the costs of production to increase because the land and labor productivity will be reduced. For example, the crops yield before climate change is 100 kg. When there is climate change such as the soil becomes acidic, the crop yield is reduced to 50 kg.

Impacts of climate change are:

1. Drop in productivity
2. Decrease in income and general welfare
3. Reduction in food security and increase food prices
4. Decrease in livelihood and welfare of farmers, followed by all citizens

10.4.1 Effect of Climate Change: The Case of Thailand Agriculture

The drought in year 2015–2016 contributes to the decline of rice production of 16% from 19.8 million tonnes to 16.5 million tonnes (Luedi 2016). However, the universities in Thailand help Thailand to develop drought-resistance seeds via genetic engineering and associated technologies. Government provides funding. However, 960,000 hectares of paddy remains unplanted due to shortage of water (Lee 2015). Thus, we can observe that the impacts of climate change are huge. We must understand it, reduce the man-made climatic changes, and adapt to nature-made climatic change.

10.5 The Causes of Climate Change in Malaysia

When there is man-made climatic change, the environment can change as well. Groundwater that becomes sour than before, soil that becomes more acidic, and increase in temperature are the examples of environmental change. This means that Malaysia is no longer planting in normal condition. Thus, Malaysia needs to make adjustments to the agriculture such as improving seeds, buying extra machines, and so forth. Malaysia is paying high costs because the indigenous technology was not developed. One major contribution to the climate change is cutting down forest to build more houses and roads. The question is why we need so many houses and roads in Malaysia. Malaysia is losing its green. If the country is losing its green, the entire ecology of agriculture may be permanently altered. During the 1950s and 1960s, the durian flowering in Malaysia is helped by the flying foxes. The flying foxes eat the nectar and pollinate the durian flower from Batu Caves to Sepang and all the villages in Malaysia. Durian trees were productively fruiting in Malaysia main durian areas. Now, in year 2017, with the cutting down of jungle and fruit trees to build houses and buildings, the flying foxes diminish and the productivity of durian decreases enormously. Without those flying foxes, the durian industry has limited production potential. The people in Malaysia have no more nice durian to eat. If the Malaysians want to eat durians, they need to import durians at high cost from Thailand. It is now difficult to grow durian in most parts of Peninsular Malaysia, with regularity of flowering season.

One can say we need to build more houses and roads when there is increase in population over time. However, that is not true. We can observe the examples of the United States, the United Kingdom, and Japan. In the United States, it is hardly any change in the road system or more roads though the population increased from 160 million people in the 1950s to 340 million in year 2017. The planning of the United States is incredibly good. The United States had been seriously noting the climatic change. Again, Malaysia should be prepared and plan for their development in the midst of the climatic changes. Malaysia must be mindful of man-made climatic changes and implement useful programs to save her food production capacity.

House construction done with unwise strategies or adverse plans may lead to flooding because the water from rain could not be absorbed into the soil once the land is cemented and the trees are gone. The retention of water may cause severe flash flooding in Malaysia. Malaysia has undergone tremendous development and urbanization. Seventy-five percent of the Malaysians live in urban (The World Bank 2016a). With this fastest urbanization, Malaysia has only about 25% of rural population that performs food production by farming as their main economic activity. This means only about 25% of Malaysia is green. Urban centers such as Kuala Lumpur, Petaling Jaya, Klang, Johor Bahru, and Penang had reduced the greens of the environment, pose disaster, and bring massive traffic congestion to urban areas in Malaysia. Malaysians do not care about green and often just look for houses rather than food from agriculture. They have not experienced food shortages or famine. They are complacent and only care for income generation. They must be reminded to reduce local man-made climatic changes.

10.6 The Sustainability of Agriculture

There are few ways to sustain agriculture. Firstly, the government must rationalize the priority in developing the agriculture and retaining farmers to stay in rural area through incentives. This either must include highland farming – Cameron Highlands – or the fishermen along the seashores. Secondly, it is the technology that matters. Technology is the application of science in daily life (Baharuddin 2013). Technology is the way of doing things that can make agriculture more productive and sustainable. The examples of technology are new varieties of crops that adapt to climate change and pest, good fertilizer, fertigation technique, and greenhouse cultivation of crops. Research and development is needed to develop new technology. Research and development (R&D) can be done in local public universities and organizations like Malaysian Agricultural Research and Development Institute (MARDI), Forest Research Institute Malaysia (FRIM), and Federal Agricultural Marketing Authority (FAMA). Research can generate new ideas to do things. Research in agriculture can be in the form of increasing yield of crops, producing crop varieties that adapt to climate change, and post-harvest product utilization. Just think of the uses of maize and soya bean in the United States. We must create new sources of wealth from Science and Technology (S&T) and R&D.

10.7 The Solutions to Sustain Agriculture

The first solution is public and private partnerships in sustaining agriculture for Malaysia. Government should have proper planning and programming on agriculture and assist the in situ farmers while developing agriculture sector. The government must not move people massively and carelessly to the urban areas. The farmers should be willing to cooperate with the government to develop the agriculture. The government and the farmers have the same objective, which is to develop agriculture and produce food to feed the nation. Government can plan good policies that can sustain the agriculture. Government can provide the basic environment for farming and give direction and focus to the direct agriculture development to produce food, feed, and fibers to the Malaysians. We must not engage in converting agricultural land to housing land at the present accelerated rate. Houses are not as basic as food for Malaysians.

Private sector refers also to the private enterprises that engage in agriculture. The private enterprises have capital, ideas, and time to develop agriculture. Capital is the man-made resource engaged in agricultural production, such as machine (Baharuddin 2013). The private enterprises need to generate ideas through funding in research to increase yield of crops and use the available capital to engage in extensive scientific agriculture farming. The private enterprises also have more time to develop the agriculture compared to government which has many more agendas for the entire nation. Thus, the private enterprises should not be dependent on government to do research and development for agriculture and rural of Malaysia.

Education also plays an important role in sustaining the agriculture. Students in the schools and universities need to appreciate the nature of agriculture. Educators should educate the students on the importance of agriculture and inspire the students to participate in planting crops and to love the nature and agriculture. The students should be instrumental in preservation of trees and the conversion of agricultural land to housing. The students with farming background can be inspired by the educators to engage in farming rather than leaving agriculture as abandoning agriculture brings harm to the entire nation and the people, especially in the congested urban environment with limited work potential and high cost of living.

Farmers should also be educated by the extension service workers that engage directly with the farmers. The extension service workers would provide guide solutions to the problems faced by the farmers. For example, climate change has negative impacts on their crops. Thus, the extension service workers would bring the issue to the experiment station, such as Malaysian Agriculture Research and Development Institute (MARDI) located near to the rural area. The experiment station would develop good crops or seeds that adapt to climate change. The extension service workers would then distribute the adapted crops or seeds to the farmers and educate the farmers to plant in their farmland. Monitoring of the crop yields is required so that improvement measure can be done if there is limitation in the crop yields and seed adaptation to climate change.

10.7.1 Ways Forward for Malaysian Agriculture Under the Stress of Climatic Change

The four sectors in the Malaysian agriculture are:

1. Crops
2. Livestock
3. Fisheries
4. Forestry

Impact assessment needs to be done by the university experts. Universities in Malaysia can have experts in climate change department, including experts to do weather forecasting and climate forecasting to prepare farmers in every state in Malaysia. We have to develop our disaster risk management capabilities.

1. Improving our natural resources management to avoid exploitation of grassland, lake, rivers, mountains, and mineral resources that seriously affect in situ climatic conditions resulting in flooding, severe droughts, and reduced rainfall

2. Reforestation and conservation to minimize effects of climatic change
3. Preparedness of all parts of the Malaysian society
4. Development of biotechnology and preservation of biodiversity

10.7.2 Short-Term Strategies and Long-Term Strategies

The short term is 1 year or one season. There is not much Malaysia can do about it. But it does not mean Malaysia do not need to do anything. Malaysia needs to prepare the alternate crops and intensify R&D on staple crops. Malaysia needs to get alternate crops that are robust to climate change. Malaysia also needs to plan and prepare to "eat less food" and develop consumption habits as a result of climate change. The short-term strategies will need more discipline and control of the markets.

The long-term strategy for Malaysia is the need to do R&D and improve food storage, processing, and packaging technologies. Through R&D, Malaysia can produce better yielding crops that suit the Malaysian soils and climates and to climatic changes. United States had even worse change in climate change before due to the geographical location and industrialization. The United States faces typhoon, cyclone, drought, snow storms, floods, and strong winds. When climate change occurred in the United States, the cost of production in agriculture increased, and the land productivity was affected. These required more conservation efforts. Over time, through R&D, varieties of plants that are resistant to climate change were developed. For example, the United States produces drought-resistant wheat through R&D. The United States is more prepared with food technology. R&D in the United States had been around US$479 billion or 2.8% of their gross domestic product (GDP) by purchasing power parity compared to Malaysia, equivalent to US$9.7 billion or 1.3% of its GDP by purchasing power parity (UNESCO 2016). This means that Malaysia must be serious with R&D investments for the future in food technology to prepare for climatic changes. Malaysian food imports totalled RM45 billion in the year 2015. This shows Malaysia had not been serious in producing food locally or finding alternate crops to prepare for future changes of climatic. Therefore, Malaysia is very vulnerable to food crisis brought about by climatic change on agriculture. Malaysia can do a few things to prepare for climatic change. The things are:

1. Malaysia must improve food technology on agriculture

   Malaysia can grow chillies and vegetables using hydroponic or fertigation techniques. The high import value of food indicates that Malaysia could face major crisis given her growing population and the negative impacts of climate change on agriculture.
2. The Malaysian society must inspire people to be farmers and not view agriculture as a poor and dirty business.
3. The Malaysian education system must be geared toward food self-sufficiency as an urgent agenda.

4. The political and government bureaucracies must put food and food security high on the budget, the plans, and research agenda.
5. Malaysia must adopt greenhouse technology and develop intensive agriculture when necessary.

10.7.3 Developing Drought-Resistant Crop

Drought-resistant crop can be developed to ensure continuous production even during drought season. For example, dry-seeded rice systems used in Bangladesh and the Philippines offer considerable economic advantages to Malaysian farmers and can ensure food security for the people. Malaysia can collaborate with the Thailand and Philippines universities in developing aerobic rice that grow in well-drained, non-puddled, and non-saturated soils. Aerobic rice can be grown in water-short irrigated lowlands (Bouman et al. 2007).

10.7.4 Development of High Productivity or High-Yielding Crop Varieties

Changes in rainfall and temperature might cause productivity losses in crops especially rice and vegetable production. For example, the drought in Malaysia can reduce the rice harvest in major granary areas in Kedah, Selangor, and Kelantan. Thus, research and development in developing high productivity rice should be intensified by local institutions including the public universities. In Malaysia, the only agriculture university is Universiti Putra Malaysia (UPM). The agenda at UPM is typical modern university, quite away from the sunset agriculture. This is not enough. Malaysia should send researchers and academicians to Thailand universities to learn rice production techniques. Thailand has good agriculture universities such as Thammasat University, Kasetsart University, and Srinakharinwirot University. Malaysia can learn the plant production and animal technology, such as from plant tissue culture, seed technology, and modern plant propagation in various institutions in Thailand under the umbrella of ASEAN and perhaps bilateral arrangements involving Thailand and Malaysia directly.

10.7.5 Land Conservation and Management

Land conservation is needed to prevent erosion and maintain fertility of soil. Research on land use is needed to understand the information on soil. Without information of soil use, Malaysia cannot do much in terms of policy to prevent erosion, planting of crops, and to maintain fertility of the soil. Land is very important to agriculture and food production. If there is no land, there is no crops. Land or soil must be fertile to encourage plant growth. Thus, research on soil nutrients is needed to understand the suitability of soil to plant growth. Farmers and related stakeholders are responsible for maintaining adequate soil fertility for good plant growth. This must be complemented by fertilizers. Human waste and animal waste such as chicken dung are important soil conditioners to maintain soil fertility for good plant growth. Understanding the ecology of plant is crucial to agricultural sustainability and address the issue of climate change. Research effort on plant pest and disease management is also left behind in Malaysia. These issues will be important to the future of foods and food security in Malaysia.

10.7.6 Fishery Sector Preparation

There are a few ways in which the fishermen can do. Firstly, the fishermen must understand the climatic and weather forecasts before going out to the sea. If the weather forecast shows that the sea is rough and stormy, the fishermen should not venture out to the sea. The weather forecasts are reported by the Malaysian broadcasting companies. Climatic changes imply a warmer ocean, which could affect fish habitat. There is a need to develop and intensify the rearing of fish in cage. The fishermen can perform some fishery-related activities such as net mending, processing fish sauce, and backyard catfish raising to improve efficiency in the fishery sector. The funding and technical advice given by the Malaysian Fisheries Development Board (Lembaga Kemajuan Ikan Malaysia) must be updated and expanded. The first step must be intensified.

10.7.7 Assessment of Risk, Vulnerability, and the Welfare of the Fishermen

The risks in fisheries include risk from climate change and risk from overcapacity (Pomeroy 2012; Heenan et al. 2015). Overcapacity could be caused by the people's poverty and the government policies to give subsidies to build bigger boats for catching more fish. Heenan et al. (2015) show that protecting the habitat for fish to grow is a way to improve resilience of the sector to climate changes and increase economic returns from fisheries. Sarawak has a good preparation way to rebuild fish stocks since 1984. The Malaysian government has allocated 4400 artificial reefs from year 1984 to year 2016 to provide breeding grounds for the fish and expanding the fish population. We need more efforts in fish cage culture, freshwater fishery, and the brackish water fish capacity. We can and must learn from all others in these issues especially from Thailand and the United States. The conservation efforts of the United States in crab and shrimp can be useful.

10.7.8 Proper Planning on the Coastal Development

There should be a proper planning on the coastal development so that any development projects near the coastline such as oil-drilling facilities and building tourism chalets would not bring erosion to the land bank and pollutes the river. Consultation between project planners and the fishermen is needed before the project is launched. Project managers must know the social, sciences, management, and environmental conservation costs brought to the fishermen because they are the most direct victim of the coastal development projects. Co-management of fisheries is needed to ensure the welfare of the stakeholders involved directly and indirectly in any development is protected. These issues also need scientific knowledge of fish and plant ecology.

10.7.9 Forest Sustainability

The government must apply laws and regulations to stop rapid or accelerated deforestation. Trees are important as they consume carbon dioxide (CO2) and supply oxygen to human beings and prevent soil erosion by holding the soil and soil water. Trees also prevent flooding that results in climate change by absorbing excess water in the soil. Thus, trees must not be simply chopped off to make way for development project and urbanization. Proper planning of land for tree plantation must be carried out to add more trees in addition to the regular reserved forests. Every person must be responsible for the environment, the trees, and the jungles. Government can encourage and inspire reforestation in the rural area. Forest provides shade for the people and absorb the sun energy. Thus, forest can reduce the negative impacts from climate change, especially overheating from sunrays. Replant of forest needs to be funded by government and the big enterprises who talk about corporate social responsibility (CSR). However, good education system should encourage voluntarism and philanthropy of the Malaysians. These must be embedded in the schools and university curriculum.

10.7.10 Strategies for Protecting Livestock

Livestock is more vulnerable to climate changes compared to crops. We can adjust crops easily to prepare for climate change. Crops for food are short-term period, ranging from 1 month to 3 months to see the production, and thus easily adjustable. On the other hand, livestock is difficult to be adjusted to climate changes. Unless

Malaysia has breeding program, the adjustments cannot be fast. Malaysia is in urgent need of livestock to provide meat and milk for the people. The self-sufficiency rate for milk in Malaysia is 7%, whereas for beef is about 24% (Department of Veterinary Malaysia 2015). Therefore, strategies for livestock especially meat and milk production alternatives are needed. Malaysia needs effective land use management and town planning for livestock to produce meat and milk to feed the people. Malaysia needs to have adequate grasslands and feedlots for the cow to graze in the field. Malaysia must not mismanage the land resources for building houses alone. We must not have all the flatlands and housing in Malaysia.

10.8 Conclusion

Agriculture is very important to Malaysia because agriculture provides food and employment to the Malaysians. However, agriculture and climate are interdependent. Plants and animals depend on the climatic factors such as the soil, sun, and water to grow. Destroys of the climatic factors may impede the growth of plants and animals. Climatic change can be caused by man-made and nature-made. Climatic change is signified by the increase in temperature, increase acidity of the soil and water, and increase of rainfalls that may have adverse effects on the plants and animals. Thus, human beings must understand the biological production process of agriculture to minimize the adverse effects of climatic change on agriculture. There are short-term and long-term strategies to sustain agriculture. The short-term strategies are making the Malaysian society more disciplined under-controlled market conditions. The examples of short-term strategies are Malaysian society must not engage in converting agricultural land to housing at the present accelerated rate, reduce the rate of cutting down forests and flattening mountains for urbanization development, and proper planning on the coastal development. The long-term strategies are Malaysia needs to do research and development (R&D) in food production technology to prepare for climate changes and also to improve her food storage, processing, and packaging technology.

References

Baharuddin AM (2013) Economics: principles, concept and applications in Malaysia. Dewan Bahasa and Pustaka, Kuala Lumpur. Published in Malay Writings

Bouman BAM, Lampayan RM, Toung TP (2007) Water management in irrigated rice: coping with water scarcity. International Rice Research Institute, Los Banos/Manila

Carvalho M, Shagar LK, Cheng N, Kanyakumari D (2016) RM45 billion for food import bill. March 15th 2016, The Star Online

Department of Statistics Malaysia (2016) Percentage share to GDP by kind of economic activity, 2015 (constant 2010 prices). Department of Statistics Malaysia, Malaysia

Department of Veterinary Malaysia (2015) Malaysia: self-sufficiency in livestock products (%), 2006-2015. Department of Veterinary Malaysia Statistics 2014/2015. Department of Veterinary Malaysia, Malaysia

Heenan A, Pomeroy R, Bell J, Munday PL, Cheung W, Logan C, Brainard R, Amri AY, Alino P, Armada N, David L, Rivera-Guieb R, Green S, Jompa J, Leonard T, Mamauag S, Parker B, Shackeroff J, Yasin Z (2015) A climate-informed, ecosystem approach to fisheries management. Mar Policy 57:182–192

Lee B (2015) Prolonged Thailand drought threatens global rice shortage. Sci Dev Net. Retrieved on July 20, 2017 from http://www.scidev.net/asia-pacific/agriculture/news/prolonged-thailand-drought-threatens-global-rice-shortage.html

Luedi J (2016). Extreme drought threatens Thailand's political stability. Global Risk Insights. Retrieved from 30 July, 2017 from https://globalriskinsights.com/2016/01/extreme-drought-threatens-thailands-political-stability/

Malaysian Palm Oil Board (2016) Oil palm planted area by state as at December 2015 (hectares). Malaysian Palm Oil Board, Malaysia

Malaysian Rubber Board (2016) Natural rubber statistics 2016. Malaysian Rubber Board, Malaysia

Mosher AT (1966) Getting agriculture moving: essentials for development and modernization. Praeger, New York

Pomeroy RS (2012) Managing overcapacity in small-scale fisheries in Southeast Asia. Mar Policy 36(2):520–527

SCB Economic Intelligence Centre (2017) Thai organic foods have healthy growth potential. Bangkok Post

The World Bank (2016a) Urban population (% of total). Retrieved on July 24, 2017 from http://data.worldbank.org/indicator/SP.URB.TOTL.IN.ZS

The World Bank (2016b) Agriculture, value added (% of GDP). Retrieved on August 5, 2017 from http://data.worldbank.org/indicator/NV.AGR.TOTL.ZS?year_high_desc=true

The World Bank (2016c) Services, etc., value added (% of GDP). Retrieved on August 5, 2017 from http://data.worldbank.org/indicator/NV.SRV.TETC.ZS

UNESCO (2016) How much does your country invest in R & D? UNESCO Institute for Statistics. Retrieved on August 5, 2017 from http://uis.unesco.org/apps/visualisations/research-and-development-spending/

United States Department of Agriculture (2016) Agriculture and food sectors and the economy. Department of Agriculture, United States

United States Department of Agriculture (2017) Palm oil production by country in 1000MT. Retrieved on July 15, 2017 from http://www.indexmundi.com/agriculture/?commodity=palm-oil

Chapter 11
Impact of Environmental Factors on Tourism Industry in Pakistan: A Study from the Last Three Decades

Hafiz Waqaz Kamran and Abdelnaser Omran

Abstract From the context of Pakistan, current study works on the key purpose to evaluate the various environmental factors and their impacts on the local tourism sector in Pakistan. For this aim there are three major dimensions of the overall environment; economic and financial indicators, social development and macroclimate changes have been selected with their key indicators based on the data sets from world development indicator (WDI). Separate standardized regression equations for each of the three factors are developed by focusing on significant statistical tests and techniques. The findings of the study reveal the fact that economic and financial indicators have a significant influence on tourism earnings, while social development and macroclimate indicators have both positive and adverse impact on the tourism industry in Pakistan. All the models are statistically significant and have provided strong evidence for the key managers to develop strategic planning for tourism industry after consideration of stated factors of the study. Additionally, this study explores the association between the key environmental factors and tourism sector in such context in a very first time. However, findings are limited to the contextual framework of Pakistan only and cannot be generalized to other economies. Finally, by adding more factors from the local environment like terrorism, the cogent for the future studies can be boosted.

Keywords Social development · Climate change · Pakistan · WDI

11.1 Introduction

Even two decades back, in the late 1990s and up to now, tourism was the largest and, among the different industry indicators, considered as among the major sources of revenues, for infrastructure and economic development as well (Awang et al. 2009;

H. W. Kamran · A. Omran (✉)
School of Economics, Finance & Banking, Universiti Utara Malaysia, Sintok, Kedah, Malaysia
e-mail: naser_elamroni@yahoo.co.uk

A. Omran, O. Schwarz-Herion (eds.), *The Impact of Climate Change on Our Life*, https://doi.org/10.1007/978-981-10-7748-7_11

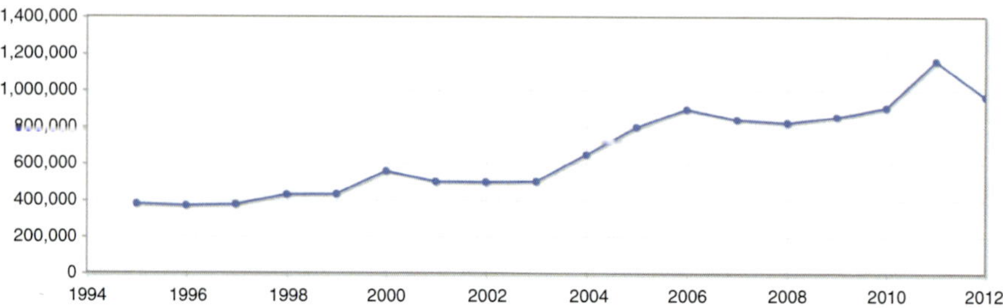

Fig. 11.1 Tourism earnings ($) over the last three decades in Pakistan (Source: https://www.indexmundi.com/facts/pakistan/international-tourism)

Chuck 1997; Narayan 2005). In the last decade, the growth of this sector is expected to be 4.6% up to 2015 with the output earning of $10.7 approximately (Hong 2008). World Tourism Organization (WTO) defined the concept of tourism as "activities of persons traveling to and staying outside their usual environment for not more than one consecutive year for leisure, business, and another purpose." The approach of research in tourism differs significantly from that of environmental sciences. Tourism industry is very much established but meanwhile with the deep research background. The traditional approach is based on the firm emphasis on the books and journal articles. However, the concept of environmental factors and tourism has got very limited attentions. Numerous studies have identified varieties of factors like sociocultural, ecological, economic, political, and technological factors in the compound business environment (Lawrence and Lorsch 1969; Milne and Ateljevic 2001; Rogerson 2002; Zhao 1994). Besides, it is the largest industry in the world economy after food, chemical, and fuels. No doubt it has a contribution toward the economic growth, but the linkage of various environmental factors with the tourism industry is yet ignored and not covered in the present body of literature. Earlier studies have covered the relationship between the tourism and terrorism (Bac et al. 2015; Baker 2014; Bowen et al. 2014; Korstanje 2015; Mukesh 2014), tourism and economic development (De Vita and Kyaw 2016; Kumar 2014; Mowforth and Munt 2015; Sharpley and Telfer 2014), tourism and external factors, and environment factors as well. Just like the other developing countries, Pakistan has a very much potential of growth of tourism industry in the South Asian region. Since the last decade, the industry of tourism is badly affected by the severity of the terrorism specifically in the northern areas (Adnan Hye and Ali Khan 2013). Pakistan has offered various places like Swat, Malam Jabba, Balakot, Ayubia, and Murree with many mountain ranges. Unfortunately, due to the earthquake in October 2005, the country has to suffer a significant loss with the decline in the number of world tourists. The major appealing area is northern which depends heavily on the tourist's income (Begum et al. 2014). The trends of tourism in the last three decades in Pakistan regarding receipts for travel items (current$) are presented with the following graph (See Fig. 11.1).

Environmental factors in any economy play a significant role for the overall situation of the country's atmosphere. Pakistan is a country which is facing various environmental issues in the last many years. Economic growth has been achieved in

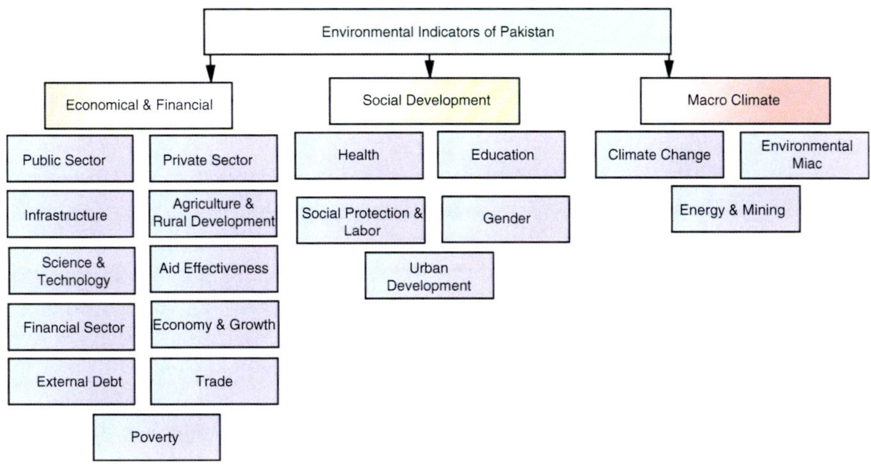

Fig. 11.2 Environmental indicators of Pakistan (model developed through personal observation, indicators from world development indicators) (Source: The authors)

the last decade with the cost of environmental issues since the majority of the production system exploits the scarce resources of the ecosystem in the country. Additionally, the dependency on the natural resources by the poor communities has increased the pressure on the environment as well (ADB 2008). By considering the previous research work, the objective of the present study is to explore the relationship between the various environmental factors and their impact on the tourism industry in Pakistan (Fig. 11.2). For achieving this objective, the overall environment of the country is divided into three broader categories:

1. Economic and financial environment
2. Social development
3. Macroclimate factors

The overall integration of these three major factors with the sub-indicators is presented in the model below:

11.2 Discussion on Existing Literature

The increasing demand for tourism with the winning number of tourists is due to the high level of attraction from the natural environment (Du Plessis et al. 2012). Such atmosphere increases the appreciation for the tourists to engage in high-quality natural experiences (Bresler 2007; Reynolds and Braithwaite 2001). Those tourists who travel to the national parks have sensory information linked to the natural areas (McCool 2006; Patterson et al. 2008). Those tourists who have positivity of such experience have total satisfaction which means a better quality of life (Borrie and Birzell 2001). For the ultimate survival of tourism in any region, focus on this element is very much integral to the sustainability of this industry (Yu and Goulden 2006).

The linkage between the tourism and environmental factors is often under discussion as many tourists decide their destination to visit based on the natural environment in that region. Such trend is very much true in various developing economies which are primarily dependent on the tourism for the economic fortune (Binns and Nel 2002; Giddy and Webb 2017). However, to develop tourism in a natural environment with the unique and exciting climate is very much challenging and the enigmatic problem with the potential threats as well (Buckley 2015; Williams and Soutar 2005). Studies of Benitez-Capistros et al. (2012, 2014) and Francisco et al. (2012) have provided a significant information for the environmental impacts in real perspective, caused by the leisure industry with some specific events as well. This matter (Giddy and Webb 2017) purely focused on the impact of environment on the tourism and adventure while considering the point of how the tourists will interact with the natural environment. The viewpoint of tourists toward the environment among the several, the most significant is the deprivation of the natural features which finally appeals the members (Giddy and Webb 2017). In the global economy, the significance of tourism industry cannot be ignored. Under the numerous topics and empirical research, the subject of tourism has covered the linkage between the tourism market, social preferences, and leisure activities. Martin and Mason (1987) have a viewpoint that those tourism activities having entertaining experience and learning can increase the demand in the mind-set of future tourists as compared to those recreational activities primarily for the rest and entertainment (Martin and Mason 1987). The study of Li et al. (2005) and Song and Li (2008) based on some statistical findings has provided the idea that earnings from the tourism with the other economic factors like exchange rate and the relative prices in the tourism terminus have a very much substantial impact on the tourism demand. Heath and Wall (1991) state that those regional communities which have got the major impact from the tourism and related factors in the recent time have taken some serious actions beyond the promotion and development. To integrate the community planning and development, more integrated strategies should be defined as the community governance, decision-making within the community based on the participation, and sustainable program as per the local setting with the sustainable tourism management (Richards and Hall 2003).

11.3 Variables, Data Collection, and Regression Models

Probably the very first study in this context by focusing on Pakistan, secondary data has been collected for the key environmental factors from data set of world development indicator. All the major indicators from WDI are integrated into three dimensions: economic and financial environment, social development, and macroclimate factors. The economic and financial indicators include the public and private sector, infrastructure, agriculture and rural development, science and technology, aid effectiveness, financial sector, economy and growth, the level of external debts, trade, and poverty in the economy. The factors for social development include the health, education, social protection and labor, gender, and urban development, while the macroclimate has considered climate changes, several factors in the environment, and finally energy and mining. After the implementation of statistical tools, only those factors have been reviewed for the final analysis which has no problem for the interdependency. For each of the major factor, separate regression models are proposed by focusing on the tourism as an explained factor of the study.

First Model: Economic and Financial

$$Tourism\$ = cons + \mathbf{b}1(MXRTRE) + \mathbf{b}2(ME2MLE) + \mathbf{b}3(ME2MESA) + \mathbf{b}4(ME2LMENA) + \mathbf{b}5(M22LMECA) + e \quad 1$$

$$Tourism\$ = cons + \mathbf{B}1(GDPPCG) + \mathbf{B}2(GDPPC) + \mathbf{B}3(NGR) + e \quad 2$$

$$Tourism\$ = cons + \boldsymbol{\beta}1(HFCE) + \boldsymbol{\beta}2(GGFCE) + e \quad 3$$

$$Tourism\$ = cons + \beta1(PPGOC) + \beta2(PPGMC) + \beta3(PPGM) + \beta4(\mathrm{PPGIBRD}) + e \quad 4$$

$$Tourism\$ = cons + \mathbf{B}1(GORPR) + \mathbf{B}2(GORCLCU) + \mathbf{B}3(NLONBPGDP) + e \quad 5$$

$$Tourism\$ = cons + \square1(ICPIAP) + \square2(CPI) + \square3(BMPGDP) + e \quad 6$$

$$Tourism\$ = cons + \boldsymbol{B}1(COCGAG) + \boldsymbol{B}2(TRIMI) + \boldsymbol{B}3(TRPTED) + e \quad 7$$

Second Model: Social Development

$$Tourism\$ = cons + \boldsymbol{\Lambda}1(ATEDU) + \boldsymbol{\Lambda}2(EILPE) + e \quad 1$$

$$Tourism\$ = cons + \boldsymbol{\lambda}1(UMLF) + \boldsymbol{\lambda}2(UF) + e \quad 2$$

$$Tourism\$ = cons + \boldsymbol{\lambda}1(PAM60 - 65) + \boldsymbol{\lambda}2(PAF60 - 65) + \boldsymbol{\lambda}3(PAM55 - 59) + e \quad 3$$

Third Model: Macroclimate

$$Tourism\$ = cons + \boldsymbol{\alpha}1(coi) + \boldsymbol{\alpha}2(fuec) + \boldsymbol{\alpha}3(coefgfckt) + e \quad \mathrm{I}$$

$$Tourism\$ = cons + \alpha(eilp) + \alpha2(fe) + \alpha3(iiewpp) + e \quad \mathrm{II}$$

11.4 Results and Discussion

Equations 1 to 5 are (Table 11.1) to express the association between tourism and key economic and financial indicators. In the very first equation 05 indicators from the economic and financial factors have been selected, which show an overall variation of 98.83 in the tourism as explained through R2. While considering all the regressors of the first equation, the modified variation in the determination of outcome factor, tourism is 97.80. The value of VIF is 7.70 for the first regression equation, an

Table 11.1 Economic and financial indicators (regression Eqs. 1–7) for tourism earnings

	1	2	3	4	5	6	7
	Tourism	Tourism	Tourism	Tourism	Tourism	Tourism	Tourism
MEXRTRE	0.00844*						
	(0.00354)						
ME2LME	32819303.9*						
	(11949744.8)						
ME2LMESA	11625005.8						
	(6363436.2)						
ME2LMENA	−11818489						
	(10144301.2)						
ME2LMECA	87577890.1***						
	(106379089)						
GDPPCG		−8401667.8					
		(8717763.9)					
GDPPC		431853.5***					
		(50822.2)					
NGR		163254792.0***					
		(38406406.)					
HFCE			0.00304***				
			(0.000388)				
GGFCE			−10422359				
			(17095130.1)				
PPGOC				0.148			
				(0.0815)			

PPGMC				1.309***			
				(0.167)			
PPGM				−0.229			
				(0.116)			
PPGIBRD				1.424**			
				(0.472)			
GORPR					3330144.8		
					(8885866.6)		
GORCLCU					0.000527**		
					(0.000139)		
NLONBPGDP					−12689702.3		
					(24851629.)		
ICPIAP						13297888.7**	
						(3997668.3)	
CPI						3336890.0***	
						(496143.8)	
BMPGDP						10517600.4*	
						(4084092.1)	
COCGAG							15776937.0**
							(4887461.4)
TRIMI							−50529484.1
							(24616577.4)
TRPTED							17781260.1***
							(3925791.7)

(continued)

Table 11.1 (continued)

	1	2	3	4	5	6	7
	Tourism	Tourism	Tourism	Tourism	Tourism	Tourism	Tourism
_cons	207067878.5***	305270940.0***	566608707.4**	67916564.2	428722945.2	−103953745.1	500160602.8***
	(39858802.8)	(40894575)	(172900921.3)	(1478537417)	(311089854.1)	(172197185.7)	(66382802.8)
f-statics	177.53	58.33	30.89	24.55	9.72	45.86	17.98
Prob > F	0.0000	0.0000	0.0000	0.0000	0.0000	0.0000	0.0000
Mean VIF	7.70	1.34	1.02	2.63	2.10	1.49	2.77
N	21	21	21	21	20	21	21
R-sq	0.983	0.911	0.774	0.86	0.646	0.89	0.76
adj. R-sq	0.978	0.896	0.749	0.825	0.579	0.871	0.718
rmse	30356170.1	65822350.3	102108479	85337888.1	134367086.7	73356992.9	108282406

Standard errors in parentheses
*$p < 0.05$, **$p < 0.01$, ***$p < 0.001$

indication of no problem for the interdependency among the explanatory factors. The time series for the selected data in the first regression equation is 21 years (1995–2015). The regressor factors MXTRE, ME2MLE, and ME2LMECA have a significant and positive impact of 0.00844, 32819303, and 87577890.1 on the tourism earnings over the last three decades. The second factor, ME2LMESA, has a negative but insignificant impact on tourism earning in the long run. The f-statistics for the first regression model is in favor for the goodness of fit for the overall model; F (5, 15) = 77.53***. All the selected factors in the first equations are from the private sector (see details in Appendix). The second regression equation has demonstrated three explanatory factors. Second regression model has shown a total variation of 91.11 in the outcome factor, tourism with the adjusted value of R2 of 89.6. (meanVIF = 1.34). The key regressors GDPPC and NGR have a significant change of 431853.5 and 163254792 in the tourism earnings. Second model is also in favor for the goodness of fit; F (3, 17) = 58.33***. The third model has regressor factors of HFCE and GGFCE. The value of F is (2, 18) = 30.89***, significant at 05% and in favor of the research hypothesis all the coefficients in the model are significantly different from zero. Total variation as explained by the third model is 77.44 and with the adjustment of the sample is 74.93 (mean VIF = 1.02). The unit change for tourism earnings from HFCE and GGFCE is 0.0030408 and −1.04e + 07, respectively. The fourth model has key regressors of PPGOC, PPGMC, PPGM, and PPGIBRD. All the selected factors for the fourth model have a significant and positive impact on the tourism earnings except PPGM which has negative coefficient −0.2291962. The mean VIF is 2.63 which indicates no problem for the correlation in the explanatory variables. The combined change in outcome factor as explained by all the predictors is 85.99 with the adjusted value of 82.49. The fifth model is also ok for the decision-making perspective as F statistics is significant with the 95% confidence level. The fifth regression model has considered GORPR, GORCLCU, and NLONBPGDP as major regressors. The model is in favor for the research hypothesis as F test has a value of 9.72 with the 95% confidence level. The percentage of response variable variation as explained by the linear model is 64.58 with the adjusted value as per the useful explanatory factors of the model is 57.94. The mean VIF is 2.10 which indicates no need for the readjustment of the model. The sixth regression model ICPIAP, CPI, and BMPGDP has been considered. All the indicators have a significant and positive impact on the value of tourism earning; the value of coefficients is 1.33e + 07, 3336890, and 1.05e + 07, respectively. The goodness of fit is at 05% level of significance with the explanatory variation by all the regressors is 89.00 and adjusted value of R2 is 87.06. The mean VIF is 1.49 demonstrating no problem for the correlation.

For the second model (Table 11.2), we have developed the three major regression equations to cover the overall social development indicators. The first regression equation for the second major indicator of overall environmental factors of ATEU and EILPE has been selected. The value of f-statistics is 47.41****, significant at 05% and in favor of the alternative hypothesis of "regression model is a good fit." The explained variation by both the key factors has 85.56 with the adjusted value of 83.76. However, the impact for the both regressors in this model is −2678632

Table 11.2 Social development indicators (regression Eq. 1) for tourism earnings

	1
	Tourism$
ATEU	−2678631.7
	(19021737.7)
EILPE	−527226487.9***
	(97588793)
_cons	3.74E + 09
	(2.29E + 09)
Mean VIF	3.39
N	19
R-sq	0.856
adj. R-sq	0.838
rmse	83648865.6

Standard errors in parentheses
*$p < 0.05$, **$p < 0.01$, ***$p < 0.001$

Table 11.3 Social development indicators (regression Eq. 2) for tourism earnings

	1
	Tourism$
UMLF	86675535.4**
	(25662524.7)
UF	−56708850.6***
	(7627299.8)
_cons	1.06791e + 09***
	(1.31E + 08)
Mean VIF	1.13
N	21
R-sq	0.757
adj. R-sq	0.73
rmse	105997725.3

Standard errors in parentheses
*$p < 0.05$, **$p < 0.01$, ***$p < 0.001$

and −5.27e + 08. Both indicators explained that when the value of ATEU and EILPE increases, it has an adverse effect on the tourism earnings. The negativity for both of these factors can be due to the deficient access to the electricity by both the urban and rural population. Meanwhile, another reason might be due to the access of both the urban and rural population with the decreasing trend. Such factors are negatively affecting the tourism earnings in Pakistan. The findings for the second model in social development are presented in Table 11.3 which has key indicators of UMLF and UF. The findings are quite consistent with the goodness of fit; no correlation issue and mean VIF are also in the tolerance level. The regression coefficients for

Table 11.4 Social development indicators (regression Eq. 03) for tourism earnings

	−1
	Tourism$
PAM60–65	−2.86240e + 09***
	(206181263.1)
PAF60–65	1.66230e + 09***
	(235135832)
PAM55–59	−688608686.5***
	(105875390.3)
_cons	5.28983e + 09***
	(766023638.5)
Mean VIF	1.17
N	21
R-sq	0.93
adj. R-sq	0.917
rmse	58624580.1

Standard errors in parentheses
* $p < 0.05$, ** $p < 0.01$, *** $p < 0.001$

UMLF and UF have a significant impact on the tourism earnings (value of coefficients, 8.67e + 07 and −5.67e + 07). The negative coefficient for the second indicator unemployment female (UF) has a major reason for the no availability of the work in the region for the females which is adversely affecting the tourism earnings in the area.

Table 11.4 represents the major indicators of social development as presented in Eq. 3 for social development factors. The findings for the very first explanatory factor have negative coefficient of −2.86240e + 09, significant at 05%; for second factor 1.66230e + 09, significant at 05%; and for third factor −688608686.5***, significant at 05%. However, the first and third explanatory factors have a negative change in the value of tourism earnings. The model is ok regarding f-statistics = 75.01, significant with the 95% confidence level. The value of R2 and adjusted R2 for this model is 92.98 and 91.74 with the best-fitted value of mean VIF of 1.17. The outcomes for a second model of macro environment are presented in Table 11.5. Three factors COI, FUEC, and COEFGFCKT are considered. The data set for these indicators is available for the last 19 years. The value of f-statistics is 17.40, significant at 01%. R2 is explaining the total variation of 77.68 with the adjustment of 04% approximately after the consideration of stated regressors of the model. The individual's coefficients are significant and negative for the FUEC and COEFGFCKT which are fossil fuel energy consumption and CO_2 emissions from gaseous consumption. Both of these indicators have a negative impact on the overall environment and climate which is adversely affecting the tourism earnings in Pakistan. The findings for the third and last regression model for the macro environmental factors are presented in Table 11.6. The key factors for this model are purely based

Table 11.5 Macroclimate indicators (regression Eq. 01) for tourism earnings

	−1
	Tourism$
COI	5442827.2
	−7865030.3
FUEC	−84829389.9***
	−18923693.1
COEFGFCKT	−1.73935e + 09*
	−695728772.5
_cons	5.20193e + 09***
	−1.13E + 09
Mean VIF	1.91
N	19
R-sq	0.777
adj. R-sq	0.732
rMSE	107418404.7

Standard errors in parentheses
$*p < 0.05$, $**p < 0.01$, $***p < 0.001$

Table 11.6 Macroclimate indicators (regression Eq. 3) for tourism earnings

	−1
	Tourism$
EILP	−397071747.7***
	−54396874
FE	36444749.5**
	−9028657.3
IIEWPP	−0.000161
	−0.0172
_cons	2.75358e + 09***
	−2.97E + 08
Mean VIF	1.28
N	13
R-sq	0.931
adj. R-sq	0.909
rmse	52998221.3

Standard errors in parentheses
$*p < 0.05$, $**p < 0.01$, $***p < 0.001$

on the energy and mining indicators. The first indicator EILP has a negative coefficient of −397071747.7*** (significant at 01%), which indicates that energy intensity level of primary energy (MJ/$2011 PPP GDP) and indicates a negative impact in the last three decades over the revenue from tourism. The third explanatory factor investment in energy with private participation also has a negative but insignificant impact on the tourism earning. The overall regression model is a good fit as the f-statistics is significant. The explained and adjusted variation in the outcome factor is finally 93% and 90% approximately.

11.5 Conclusion

The present study focuses on the overall environmental indicators which have ultimate influence on the tourism earnings from the context of Pakistan. The key indicators have been divided into three broad categories: economic and financial environment, social development, and finally macroclimate changes. All the main indicators have sub-factors for which data is collected from WDI database. A strong, significant, and positive impact of economic and financial indicators has been found on the tourism earnings in the last three decades, while the other two indicators have presented a mixed trend in tourism revenue in the long run. The study itself is a useful contribution to the existing literature by the following principal reasons.

1. It is focusing on the overall dimensions of country's environment which include three major aspects (economic and financial environment, social development, and major climate) which are not addressed up to now in any study.
2. Demonstrating the very first relationship with the range of indicators to tourism.
3. The study is also contributing with the strong findings to provide the guidelines to the main representatives of the state to consider the first and such meaningful association between tourism and all the environmental indicators.

Apart from this, it is also recommended for the country management to develop appropriate strategies which can attract more and more tourists and visitors in the northern region specifically. Besides the above contribution and recommendations, some issues and gaps need to be resolved and covered. At first, the data set consists of maximum last 21 years from the context of Pakistan. If the time series of data configured to expand, it will most probably provide some more significant findings. At second, Pakistan is in high threat of terrorism which has adversely affected the tourism earnings in the northern areas. So, the factor of terrorism must also be considered to cover the existing gap in the study. Finally, a simple groupwise regression models have been applied, but this technique is not very much advance, so applying the better statistical and time series technique can provide the better findings comparatively to the present one.

Appendix: Glossary and Abbreviation

Sr. no	Abbreviation	Name of variable
01	MEXRTRE	Merchandise exports by the reporting economy (current US$)
02	ME2LME	Merchandise exports to low- and middle-income economies in sub-Saharan Africa (% of total merchandise exports)
03	ME2LMESA	Merchandise exports to low- and middle-income economies in South Asia (% of total merchandise exports)
04	ME2LME and NA	Merchandise exports to low- and middle-income economies in Middle East and North Africa (% of total merchandise exports)
05	ME2LME and CA	Merchandise exports to low- and middle-income economies in Europe and Central Asia (% of total merchandise exports)
06	GDPPCG	GDP per capita growth (annual %)
07	GDPPC	GDP per capita (current US$)
08	NGR	Natural gas rents (% of GDP)
09	HFCE	Household final consumption expenditure, etc. (current US$)
10	GGFCE	General government final consumption expenditure (% of GDP)
11	PPGOC	PPG, official creditors (AMT, current US$)
12	PPGMC	PPG, multilateral concessional (AMT, current US$)
13	PPGM	PPG, multilateral (AMT, current US$)
14	PPG,IBRD	PPG, IBRD (AMT, current US$)
15	GORPR	Grants and other revenue (% of revenue)
16	GORCLCU	Grants and other revenue (current LCU)
17	NLONBPGDP	Net lending (+) / net borrowing (−) (% of GDP)
18	ICPIAP	Inflation, consumer prices (annual %)
19	CPI	Consumer price index (2010 = 100)
20	BMPGDP	Broad money (% of GDP)
21	COCGAG	Claims on central government (annual growth as % of broad money)
22	TRIMI	Total reserves in months of imports
23	TRPTED	Total reserves (% of total external debt)
24	ATEU	Access to electricity, urban (% of urban population)
25	EILPE	Energy intensity level of primary energy (MJ/$2011 PPP GDP)
26	UMLF	Unemployment, male (% of male labor force) (modeled ILO estimate)
27	UF	Unemployment, female (% of female labor force) (modeled ILO estimate)
28	PAM60–65	Population ages 60–64, male (% of male population)
29	PAF60–65	Population ages 60–64, female (% of female population)
30	PAM55–59	Population ages 55–59, male (% of male population)
31	COI	CO2 intensity (kg per kg of oil equivalent energy use)
32	FUEC	Fossil fuel energy consumption (% of total)
33	COEFGFCKT	CO2 emissions from gaseous fuel consumption (kt)
34	EILP	Energy intensity level of primary energy (MJ/$2011 PPP GDP)
35	FE	Fuel exports (% of merchandise exports)
36	IIEWPP	Investment in energy with private participation

References

ADB (2008) Islamic Republic of Pakistan country environment analysis. ADB, South Asia, p 58
Adnan Hye QM, Ali Khan RE (2013) Tourism-led growth hypothesis: a case study of Pakistan. Asia Pac J Tour Res 18(4):303–313
Awang KW, Hassan WMW, Zahari MSM (2009) Tourism development: a geographical perspective. Asian Soc Sci 5(5):67–76
Bac DP, Bugnar NG, Mester LE (2015) Terrorism and its impacts on the tourism industry. Revista Română de Geografie Politică 1:5–11
Baker DMA (2014) The effects of terrorism on the travel and tourism industry. Int J Relig Tour Pilgrimage 2(1):1–11
Begum N, Malik AA, Shahzad S (2014) Prevalence of psychological problems among survivors of the earthquake in northern areas of Pakistan. Asian J Manage Sci Educ 3(3):49–57
Benítez-Capistrós F, Hugé J, Koedam N (2012) Environmental impacts on the Galapagos Islands: Identification of interactions, perceptions, and steps forward
Benitez-Capistros F, Hugé J, Koedam N (2014) Environmental impacts on the Galapagos Islands: identification of interactions, perceptions, and steps ahead. Ecol Indic 38:113–123
Binns T, Nel E (2002) Tourism as a local development strategy in South Africa. Geogr J 168(3):235–247
Borrie WT, Birzell RM (2001) Approaches to measuring the quality of the wilderness experience. Visitor use density and wilderness experience: proceedings, 2000 June 13. In: Proceedings RMRS-P-20, Missoula, p 29
Bowen C, Fidgeon P, Page SJ (2014) Maritime tourism and terrorism: customer perceptions of the potential terrorist threat to cruise shipping. Curr Issue Tour 17(7):610–639
Bresler N (2007) Wildlife tourism: creating memorable and differentiated experiences. Acta Acad 39(3):165–182
Buckley RC (2015) Adventure thrills are addictive. Front Psychol 6:1915. https://doi.org/10.3389/fpsyg.2015.01915
Chuck Y (1997). International tourism: a global perspective. World Tourism Organization (WTO) Publications
De Vita G, Kyaw KS (2016) Tourism development and growth. Ann Tour Res 60:23–26
Du Plessis ML, van der Merwe P, Saayman M (2012) Environmental factors are affecting tourists' experience in South African national parks. Afr J Bus Manage 6(8):2911
Francisco B-C, Hugé J, Koedam, N (2012) Environmental impacts on the Galapagos Islands: identification of interactions, perceptions, and steps forward. Book of Abstracts, pp 1–149
Giddy, JK, Webb, NL (2017) Environmental attitudes and adventure tourism motivations. GeoJournal, pp 1–13
Heath E, Wall G (1991) Marketing tourism destinations: a strategic planning approach. Wiley, New York
Hong SW-C (2008) Competitiveness in the tourism sector: a comprehensive approach from Economic and Management points. Springer, Heidelberg
Korstanje ME (2015) The spirit of terrorism: tourism, unionization and terrorism. Pasos 13(1):239–250
Kumar RR (2014) Exploring the role of technology, tourism and financial development: an empirical study of Vietnam. Qual Quant 48(5):2881–2898
Lawrence PR, Lorsch JW (1969) Developing organizations: diagnosis and action. Addison-Wesley Publishing, Reading
Li G, Song H, Witt SF (2005) Recent developments in econometric modeling and forecasting. J Travel Res 44(1):82–99
Martin W, Mason S (1987) Social trends and tourism futures. Tour Manage 8(2):112–114
McCool SF (2006) Managing for visitor experiences in protected areas: promising opportunities and fundamental challenges. Parks 16(2):3–9

Milne S, Ateljevic I (2001) Tourism, economic development, and the global-local nexus: theory embracing complexity. Tour Geogr 3(4):369–393
Mowforth M, Munt I (2015) Tourism and sustainability: development, globalization and new tourism in the third world. Routledge, London
Mukesh R (2014) Terrorism terrorizes tourism: Indian tourism effacing myths? Int J Saf Secur Tour 5:26–39
Narayan PK (2005) The structure of tourist expenditure in Fiji: evidence from unit root structural break tests. Appl Econ 37(10):1157–1161
Patterson TM, Niccolucci V, Marchettini N (2008) Adaptive environmental management of tourism in the Province of Siena, Italy using the ecological footprint. J Environ Manage 86(2):407–418
Reynolds PC, Braithwaite D (2001) Towards a conceptual framework for wildlife tourism. Tour Manage 22(1):31–42
Richards G, Hall D (2003) Tourism and sustainable community development, vol 7. Psychology Press, London
Rogerson CM (2002) Tourism-led local economic development: the South African experience. Paper presented at the Urban Forum
Sharpley R, Telfer DJ (2014) Tourism and development: concepts and issues, 2nd edn. Channel View Publications, London
Song H, Li G (2008) Tourism demand modelling and forecasting – a review of recent research. Tour Manage 29(2):203–220
Williams P, Soutar G (2005) Close to the "edge": critical issues for adventure tourism operators. Asia Pac J Tour Res 10(3):247–261
Yu L, Goulden M (2006) A comparative analysis of international tourists' satisfaction in Mongolia. Tour Manage 27(6):1331–1342
Zhao X (1994) Barter tourism along the China-Russia border. Ann Tour Res 21(2):401–403

Chapter 12
Anthropogenic Climate Change and Countermeasures: Chances and Risks of Weather Modification Techniques and Climate Engineering (CE)

Odile Schwarz-Herion

Abstract There are different measures to tackle the challenge of Climate Change. This includes weather modification techniques and climate engineering (CE). Especially the application of CE is very controversial. In this connection, one has to differentiate between carbon dioxide removal (CDR) techniques and solar radiation management (SRM) techniques. According to climate experts, CDR techniques – except from the controversial CO_2 sequestration and some other risky CDR techniques – interfere less aggressively into the natural environment than SRM techniques. Additionally, CDR puts on the causes of Climate Change, whereas SRM merely treats the symptoms. Apart from possible negative side effects, environmental modification techniques and CE might also be abused and have already been abused for covert weather warfare or terrorism by the deliberate aggravation or creation of extreme weather patterns and natural disasters like droughts, blizzards, floods, and storms with the intention to cause property damages, health problems, injuries, or even fatalities in certain areas or nations to harm the enemy. Although weather and climate modification for military or other hostile purposes are expressly prohibited by the UN's 1977 ENMOD Convention, this UN convention is repeatedly circumvented. In fact – additionally to the inadvertent anthropogenic climate change – deliberate anthropogenic Climate Change seems to be feasible nowadays by the use of existing technology, allowing a range of possibilities for targeted large-scale anthropogenic modification and manipulation of the weather and possibly even the climate.

Keywords Anthropogenic Climate Change · Weather modification techniques · Climate engineering (CE) · Solar radiation management (SRM) · Carbon dioxide removal (CDR) · UN ENMOD Convention · Covert weather warfare

---

O. Schwarz-Herion (✉)
Dres. Shwarz-Herion KG Detective Agency, Ettlingen, Germany
e-mail: EOKK.Black@t-online.de

A. Omran, O. Schwarz-Herion (eds.), *The Impact of Climate Change on Our Life*, https://doi.org/10.1007/978-981-10-7748-7_12

12.1 Introduction

The accumulation of severe natural disasters is obvious since many decades already: The 2004 earthquake of magnitude 9.0 in Indonesia triggering tsunamis affecting Indonesia, other Asian countries, and the east coasts of Africa in which "…an estimated 250, 000 persons perished…" (Srinivas 2015); *Hurricane Katrina* from 2005, killing 1577 persons in Louisiana, and hundreds of persons in other US states (Knabb et al. 2011); *Hurricane Sandy* from 2012, causing 147 recorded direct deaths[1] across the Atlantic basin (Blake et al. 2013); and the *Kelantan flood* from 2013 in Malaysia (see Chap. 2) are only some widely known weather disasters in the more recent past.

Actually, there have been other, very severe weather events already many decades ago. This did not only include the previous high floods in Malaysia between 1926 and 2004 (see Chap. 2) but also other dramatic weather disasters registered only recently by the World Meteorological Association (WMO). These disasters included a hailstorm in Honan Province, Nanking, China, killing 200 people back in 1932, furthermore, the "Great Bhola" cyclone in Bangladesh killing 300,000 persons in 1970, a fatal lightning strike killing 21 persons in Zimbabwe in 1975, a hailstorm killing 246 persons in India with hailstones "as large as 'goose eggs and oranges' and cricket balls" in 1988 (WMO 2017a), a tornado in Bangladesh leaving 1300 persons dead, and a lightning strike in Egypt in 1994, putting a tank with kerosene and diesel into fire, killing 469 persons (WMO 2017b). These fatal natural disasters happened at a time when Climate Change had still no or only little coverage in the mass media.

While little can be done against the natural factors contributing to Climate Change, it might be possible to do something against the anthropogenic factors. Some of the aforementioned severe weather disasters might have been enhanced by anthropogenic factors, eventually by greenhouse gas emissions as alleged, e.g., by the IPCC. Another anthropogenic factor which might contribute to Climate Change but which is often entirely ignored in scientific publications is the deliberate manipulation of the weather and the climate by the covert use of *environmental modification techniques* and *climate engineering (CE)* for scientific, military, political, or economic reasons. The chapter at hand is supposed to fill this gap by including this factor in the discussion of anthropogenic Climate Change and in the list of recommendations for strategies against Climate Change which have to take into account the most probable causes of Climate Change.

[1] Direct deaths are those fatalities caused by the storm itself, e.g., by drowning in storm surge, rough seas, rip currents and fresh water floods, by lightning, house fires or wind-related events like collapsing structures whereas deaths indirectly caused by the storm include, for example, electrocutions from drowned powerlines, heart attacks as indirect consequences of the storm (Blake et al. 2013).

12.2 Climate Change: Possible Causes and Consequences

12.2.1 Possible Causes

Since 1900, the Earth's average surface temperature has increased by about 0.8 °C; much of this increase took place in the 1970s. The best evidence for global warming comes from widespread thermometer records, partially traced back to the nineteenth century (The Royal Society 2014).

Climate Change can be caused by natural factors and anthropogenic factors. Natural factors include variations in the Sun's output and in the Earth's orbit around the sun, volcanic eruptions, and internal fluctuations in the climate system like El Niño and La Niña, and – especially for the Northern hemisphere – the arctic oscillation (Royal Society 2014; Schwarz-Herion 2005, with further references). Calculations based on climate models have simulated the effects of natural causes without any human influence on global temperature. According to these simulations, there was little warming or even a slight cooling over the course of the twentieth century, which might lead to a new Ice Age. Only if human influences on the composition of the atmosphere are taken into account in model simulations, "...the resulting temperature changes are consistent with observed changes..." (The Royal Society 2014).

The so-called greenhouse effect plays a crucial role for global warming. The *Royal Society* describes it as follows:

> ...The Sun is the primary source for the Earth's climate. Some of the incoming sunlight is reflected directly back into space, especially by bright surfaces such as ice and clouds, and the rest is absorbed by the surface and the atmosphere. Much of this absorbed solar energy is re-emitted as heat (longwave or infrared radiation). The atmosphere, in turn, absorbs and re-radiates heat, some of which escapes to space. Any disturbance to this balance of incoming and outcoming energy will affect the climate.... (The Royal Society 2014)

Greenhouse gases (GHG) which are naturally occurring in the atmosphere like water vapor (H_2O), carbon dioxide (CO_2), ozone (O_3), methane (CH_4), and nitrous gases (NO_x) "...work like the glass sheet of a greenhouse, allowing the short-wave UV radiation of the sun to pass almost unhindered, whereas the long-wave infrared radiation (warming radiation) is mainly absorbed by the surface of the earth..." (Schwarz-Herion 2015, with further references), stabilizing the average temperature at ca. 15 °C on Earth (Schwarz-Herion 2015, with further references). In fact, the greenhouse effect is the base for all life on Earth (The Royal Society 2014). A significant increase of GHG in the atmosphere by natural or anthropogenic (human-made) influence "...can disturb the radiation equilibrium in the earth's atmosphere, which might result into a global increase in the surface temperature of the earth..." (Schwarz-Herion 2005, 2015, with further references).

Climate research shows that there has been an increase in the global surface temperature of the Earth since ca. 1500 which has further increased since 1800. So, global warming is irrefutable. Controversial is only the question whether this global warming has anthropogenic or natural causes. The volcanic eruptions, for example, account to only 1% of increase in CO_2 which might be an indicator that the increase in the Earth's surface temperature might be mainly caused by human activities (The Royal Society 2014).

The human factors can be subdivided into inadvertent and deliberate human factors. Inadvertent human factors, for example, are human activities in industry, street traffic, and households causing GHG emissions from burning fossil fuels (Uther 2014; see also Chap. 1). The depletion of natural sinks like forests by large-scale deforestation can enhance the amount of CO_2 in the atmosphere and thus can further contribute to Climate Change.

Deliberate human factors are the covert use of weather modification techniques and climate engineering (CE). These techniques exist since several decades already. Some of them have already been covertly applied for civil or military purposes as to be shown later on in this chapter.

12.2.2 Climate Change: Possible Consequences

According to climate experts, anthropogenic Climate Change can cause weather disasters and extreme weather patterns (NOAA 2015; IPCC 2014; The Royal Society 2014). Since "…the earth's lower atmosphere is becoming warmer and moister…" (The Royal Society 2014), this provides more energy for storms and certain severe weather events (The Royal Society 2014). Furthermore, "…the melting of the polar ice-caps and the Greenland ice-sheet and the thermal expansion of the water lead to an increase in the sea level…heavy rainfalls and snowfall events… and heat waves are generally becoming more frequent…" (The Royal Society 2014). Beyond that, Climate Change increases CO_2 levels in the oceans, thus leading to acidification of the oceans (The Royal Society 2014).

Many other direct and indirect consequences of Climate Change have already been addressed (see Chap. 1 and see above in the introduction). Some of these effects can also be produced and enhanced by deliberate human activities, i.e., by the use of weather modification techniques or CE. These technologies will be explained in this chapter in a detailed way later on.

12.3 Countermeasures Against Climate Change

12.3.1 Conventional Measures to Reduce or Prevent CO_2 Emissions

Conventional countermeasures against Climate Change, based on the assumption that Climate Change was due to GHG emissions from industry, households, and traffic, include the use of low-consumption motor vehicles with conventional drives and those with alternative drives reducing CO_2 emissions significantly (hybrid cars and electric cars running on renewable energies) or would even prevent them completely (fuel cell cars). In the field of energy production, power plants based on fossil energy sources (coal, fuel, and natural gas) should gradually be replaced by power plants based on renewable energies like solar, wind, and water energy. Suitable measures should be taken to reduce energy consumption in both industrial plants and private households, e.g., by thermal insulation, solar roofs, and ecological industrial or household appliances, respectively.

Nonetheless, it can be problematic in democratic states to implement appropriate measures in the framework of a free market economy without restricting the liberty of private individuals or the economic agents too much. Therefore, the trade with emission certificates with the objective to confer "...marketable rights for the use of the environment which would entitle someone to the emission of a certain amount of, e.g., CO_2 within and outside of certain spatial and temporal boundaries..." seemed to be a reasonable compromise (Schwarz-Herion 2005, with further references). This trade has, however, suffered a reputational loss due to "...large-scale fraud schemes in trading CO_2 emission certificates via non-EU countries as transit countries, targeted at circumventing EU transaction fees..." (Schwarz-Herion 2015, with further references), a fraudulent system into which also *Deutsche Bank* has been involved (Schwarz-Herion 2015, with further references).

12.3.2 Special Technological Mitigation Measures: Weather Modification Technologies

Early weather modification techniques were already capable of the artificial creation or the enhancement of rainfall (Vidal and Weinstein 2001; Pook 2001). In the nineteenth century and at the beginning of the twentieth century, there existed already some elaborated weather modification patents to prevent droughts and crop failures.

Let us have a look at two examples for early "rainmaking patents":

1. *United States Patent 0462795 – Method Of Producing Rain-Fall*, invented by *Louis Gathmann* of Chicago, Illinois:

   "...a method of producing a rain-fall...to produce a condensation in the upper regions of the atmospheric air in such quantities that a cloud will be formed from which a rain-fall will be precipitated...various means could be employed...to suddenly chill the atmosphere by rapid evaporation..." (Gathman 1891) by "...liquefied carbonic-acid...which is thrown or shot into the upper regions of the atmosphere and there exploded by a time-fuse. A balloon... employed to elevate the shell or casing containing the liquefied carbonic-acid gas, and the explosion to liberate the gas could be made by an electric current controlled by persons upon the earth ...." (Gathmann 1891)
2. *United States Patent 1,103,490 RAIN-MAKER*, invented by *James M. Cordray*, United States, filed on August 6, 1913, and patented on July 14, 1914, a method of "...producing rain...in those sections of the world where rainfalls occur at

widely separated intervals to produce rain during ...dry season… in other sections of the world during exceptionally prolonged droughts…" (Cordray 1914). The object of this invention was "...to accomplish this by supplying moisture, heat, and nitrogen to the air by means of explosives raised above the earth's surface, by balloons, kites, or other suitable devices…" (Cordray 1914), using balloons to raise "…water in tanks…provided with a heater by…which the water contained therein may be turned to steam to moisten and warm the atmosphere… electrically treating the atmosphere …" (Cordray 1914).

Interestingly, Gathmann's patent from 1891 was targeted at ***cooling*** the atmosphere, whereas Cordray's patent from 1914 strived for ***warming*** the atmosphere. This shows that weather modification technology was already rather advanced and versatile over 100 years ago.

In the 1940s, the technology of "cloud seeding" was implemented by the chemist and Nobel Laureate winner *Irving Langmuir*. In this process, a cloud was "vaccinated" by spraying silver iodide into the sky, thus producing artificial precipitation. A series of similar weather modification projects for the production of rain and snowfall followed in subsequent years and decades (Uther 2014, with further references; Rich 2011; Coles 2010; Fleming 2007).

Technology for the modification of storms is also available since several decades. Early examples include the project "Stormfury" which was carried out between 1962 and 1983. This project aimed at "vaccinating" a hurricane during its genesis to resolve the storm this way (Uther 2014, with further references). Since the mid-1990s, there exist patents for tropical cyclone disruption. This includes *Patent US 5441200 A* whose method is described in the following way:

> …a chemical which allows water to chemically join its crystalline lattice is applied to the eye wall of a tropical cyclone to initiate a self-destructive catalyzing effect. If applied in powdered form to the upper, center portions of the eye wall, the effect will be greater. Water vapor within the eye wall chemically joins the lattice of the chemical. These larger molecules will also develop through collision and coalesce. Now the vapor of the eye wall is heavier and will spin outwards from Centrifugal Force. As a result of the larger eye, barometric pressure in the eye increases, wind speed slows, and the storm surge decreases to minimal proportions. (Rovella 1995)

Meanwhile, there is a whole range of technologies for weather modification to mitigate the consequences of Climate Change. Since the early 1960s, the private company *Weather Modification Incorporated* has offered a professional "weather modification service" to insurance companies, water resource management groups, and federal and state government research organizations specializing in weather modification, cloud seeding, rain enhancement, and fog dispersion (Rich 2011). Actually, this company still exists and is officially advertising its services to satisfy the need for "…atmospheric necessities such as water resource management and environmental quality monitoring…" (Weather Modification Inc. 2017). These rainmaking technologies can help to mitigate or avoid the worst consequences of droughts like drought-related fatalities of humans and animals, water scarcity, and crop failures.

Conversely, the US company *Dyn-O-Mat* wants to prevent rainfall. For this purpose, it has developed a special gel, which is scattered by airplanes and absorbs so much air humidity that cloud formation does **not** occur (BBB Better Business Bureau 2017; Odenwald 2007; Trademarkia 2003; Eastwood 2001; Siddle 2001). Technologies for the mitigation and the prevention of hailstorms have become more frequent due to Climate Change (Odenwald 2007). In sum, there exist already civil programs in 30 countries to modify the weather since 2007 (Odenwald 2007).

12.3.3 Innovative High-Tech Solutions: Geo-Engineering Aka Climate Engineering (CE)

"If somebody is driving an SUV, I would be against deploying geo-engineering. If there is a coal-fired power plant still spewing carbon dioxide into the atmosphere, I would be against geo-engineering. We should think of geo-engineering only as a parachute…It's something you desperately hope you never need" (Caldeira 2007).

This statement shows how skeptical people had still been 10 years ago toward geo-engineering. While one could read in a 2008 article in the *Foreign Affairs Magazine* that "…nearly the entire community of geo-engineering scientists could fit comfortably in a single university room…" (Uther 2014), this has changed enormously since then. The academic literature on geo-engineering, also known as climate engineering (CE), has skyrocketed over the past 5 years.

What is geo-engineering/CE? There are many different definitions for geo-engineering/CE. According to the 2009 Royal Society Report on CE, the term is defined as "…the deliberate large-scale intervention in the Earth's Climate System, in order to moderate global warming…"(Thornes and Pope 2014, with further references). Some scientists and scholars prefer the term "climate engineering" (CE) as more precise and clearer, since this term refers to interventions into the Earth's climate system (Schwarz-Herion 2015, with further references; Uther 2014, with further references). This term is also used in this chapter.

One has to distinguish between CE and weather modification. The difference between these technologies is the spatial and temporal scope as well as the scope of the respective measures. While climatic changes take place over the periods of decades and centuries and have an over-regional effect, accordingly requiring enduring and large-scale countermeasures like CE, weather modifications or manipulations produce only short-term, spatially limited changes in the atmosphere, i.e., weather changes (Thornes and Pope 2014).

Furthermore, CE is "…distinct from Climate Change mitigation and adaptation. Mitigation involves strategies to reduce anthropogenic emissions of greenhouse gases, for example through the transition to a low carbon economy. Adaptation aims to decrease resilience…towards the effects of Climate Change, for example, by improving flood defenses..." (Thornes and Pope 2014, with further references).

Nevertheless, "…the Paris Agreement has high aspirations, and its overall goal – to 'hold the increase in the global average temperature to well below 2 °C above pre-industrial levels and to pursue efforts to limit the temperature increase to 1.5 °C above pre-industrial levels' – can't help but put geo-engineering firmly in the frame" (Shepherd 2016; Faber 2015). Therefore, CE is often seen as "ultima ratio" to save the world from the consequences of Climate Change and a possible climate collapse (Faber 2015).

A large-scale study by the German sociologist Stephanie Uther deals with the discourse, i.e., the controversial discussion about CE in Great Britain and Germany in science, the media, and politics. Substantial, recurring arguments are being worked out both for and against CE (Uther 2014). It is typical for the development of the discussion that it initially takes place only among scientists, then spreads to the media, and finally to politics, whereby the discourse levels can overlap (Uther 2014).

An alleged "lack of alternatives" is an especially popular argument for the exploration of CE. The uncertainties of the effectiveness of CE, i.e., the question if they can have any effect on the global average temperature, however, speak against this argument (Uther 2014). Some CE opponents even argue that the knowledge of the CE option alone could drive humanity to continue to emit CO_2 in the future (the so-called "moral hazard" argument). Wasting time and resources for CE research indeed prevents other options, e.g., renewable energies, from being researched (Uther 2014, with further references). Others argue that especially the existing knowledge gaps concerning CE require detailed research including the implementation of CE to be able to assess its risks and side effects more thoroughly (Uther 2014).

Another argument for the alleged necessity of CE research is "Arming the Future," alleging that humanity needs to explore CE options in order to give future generations an optimal basis for decision-making (Uther 2014, with reference to Ott 2011). This can be countered by the argument that the use of some CE technologies can entail risks for future generations since the consequences of CE deployment can be more serious than the actual consequences of Climate Change by natural factors and by inadvertent human activities like industrial activities, street traffic, etc.. Thus, some argue that it was irresponsible to use such technologies and that CE was a negative example for the blind faith in the technological progress of mankind (Uther 2014).

In addition, the use of CE might lead to political conflicts on international level (Uther 2014, with further references). In 2010, a tentative CE agreement was established at the *International Conference on Conventional Biological Diversity (CBD)*, as well as a *de facto moratorium* which instructed the member countries to do "… no climate-related geo-engineering activities that may affect biodiversity, until there is an adequate scientific basis on which to justify such activities…" (Uther 2014, with further references). This prohibition, however, concerned only the scope of the CBD and was not yet legally binding (Uther 2014, with further references). Two years later, the CBD once again turned to CE and issued an assessment of legal and political regulatory requirements. This shows the high political importance of the CBD decision, even if the *moratorium* is not yet legally binding. On October 18, 2013, the contracting parties to the *London Protocol* decided to exacerbate former regulations on marine geo-engineering and to ban the commercial sewage of the seas, except for legitimate scientific research projects which are distinct from commercial activities (Uther 2014, with further references).

The two basic techniques of CE are *carbon dioxide reduction (CDR) techniques* removing CO_2 from the atmosphere, such as "…land use management to protect or enhance land carbon sinks, the use of biomass for carbon sequestration as well as carbon neutral energy source, acceleration of natural geological weathering processes that remove CO_2 from the atmosphere…" (Schwarz-Herion 2015, with reference to The Royal Society 2009) and *Solar Radiation Management (SRM)* techniques "…that reflect a small percentage of the sun's light and heat back into space…" (Schwarz-Herion 2015, with reference to The Royal Society 2009). SRM can be differentiated by (a) space-based reflectors, (b) technical production of clouds, and (c) introducing large amounts of sulfates into the atmosphere (Schwarz-Herion 2015, with reference to The Royal Society 2009). Furthermore, "…SRM schemes decrease the effects of Climate Change by reducing the amount of energy within the earth system by reflecting a proportion of solar radiation back to space…" (Thornes and Pope 2014).

Climate experts and critics of CE regard CDR – with some exceptions like the controversial CO_2 sequestration – generally as less problematic than SRM, because CDR "…fights the cause, whereas SRM only treats the symptoms…" (Schwarz-Herion 2015, with reference to Ott 2011) and furthermore since "…CDR meshes less deeply into the nature than SRM which can bring about undesired Climate Change for humans and nature under certain circumstances…"(Schwarz-Herion 2015, with reference to Ott 2011).

A particularly controversial issue regarding SRM is the introduction of sulfate aerosols into the atmosphere, because "…this would nullify what has been done some decades ago by using desulfurization technologies: the protection against acid rain…" (Schwarz-Herion 2015). Sulfate aerosols containing sulfur dioxide (SO_2) do not only destroy the essential Vitamins B1 and B12 (Belitz 1999) but are also poisonous and can cause respiratory paralysis – in contrary to CO_2 as essential non-toxic gas (CIA 2002).

Allegedly, CO_2 is especially suitable for the application of CDR, because its extraction is easier due to its supposed high concentration (Thornes and Pope 2014). CDR measures encompass chemical, physical, and biological techniques for removing CO_2 from the atmosphere which can be implemented on land or in the ocean (Uther 2014, with further references).

The technology-based lowering of the ocean's flow to transport colder water to the Earth's surface and to stimulate marine carbon storage is one of the chemical and physical methods of CDR (Uther 2014, with further references). Another method known as "enhanced weathering" involves chemical processes for CO_2 sequestration in which natural weathering processes are accelerated, for example, to dissolve silicate rocks (Uther 2014, with further references).

Other land-based measures can be summarized as "air capture." In industrial processes, CO_2 is filtered out of the ambient air and pure CO_2 is left behind as a residue. Because of its futuristic appearance, the concept of "artificial trees" attained prominence: CO_2 can be absorbed from the air and stored, using chemical sorption agents (Uther 2014, with further references). Nevertheless, artificial trees are criticized by experts for their enormous environmental and economic costs and thus do not really pay off (Biello 2009).

Biological CDR measures include ocean fertilization with macro- or micronutrients (e.g., iron, phosphorus, nitrogen). In nutrient-poor marine regions, fertilization should strengthen the process of photosynthesis in marine phytoplankton. Through initial studies such as *LOHAFEX*, an Ocean Iron Fertilization experiment, this topic became public and was surprisingly stopped by the *German Federal Ministry of Education and Research* after environmental organizations, citizens' initiatives, scientists, and the *Federal Ministry of the Environment (BMU)* regarded this as a violation of international agreements about the overfertilization of the oceans and warned against the grave impact on the marine ecosystem. In the media, this experiment made headlines, among others, in the renowned German magazine *Der Spiegel* which warned against "incalculable interventions in the oceans" (Uther 2014, with reference to Der Spiegel 2009).

A textbook example for the aforementioned CDR technique in the ocean is *Patent US 8033879*, also known as *US 2009022716 B2-Biophysical geo-engineering compositions and methods*, invented by and assigned to *Kal K Lambert* (Lambert 2011). The inventor describes this patent in the following way:

> …compositions, methods and apparatus for biological and physical geoengineering…inorganic particles...for dispersal on a body of water, the compositions having several properties…including a positive buoyancy and…a sustained-release matrix for delivery of iron, calcium, magnesium, zinc, copper, molybdenum, manganese, cobalt, boron, selenium, vanadium, chromium, nickel, sulfur, nitrogen, phosphorus, silicon…a light-reflective surface for increasing planetary albedo[2] when dispersed over large surfaces…to…increase yields for pelagic aquaculture…increase carbon sequestration…provide immediate relief from global warming by directly increasing surface albedo and indirectly by increasing cloud nucleation activity…comprising a mixture of inorganic salts and binders (such as a glass or ceramic) effective for increasing the growth of primary aquatic producers such as coccolithophorids, diatoms, silicoflagellates, dinoflagellates, and microalgae…To validate resultant carbon fixation and sequestration, a vertical spar buoy is provided…designed to resist wave motion while supporting a stable long-term platform and submerged spaceframe with laser particle counters suspended below the 100 year horizon…. (Lambert 2011)

The inventor stresses that "...about 30% of the carbon dioxide emissions from fossil fuel combustion" (Lambert 2011) is being absorbed in the oceans annually, that "…higher rates of increase in atmospheric CO2 tend to occur in El Niño years…" (Lambert 2011), and that "...atmospheric CO2 will continue to rise even if fossil fuel combustion is stopped tomorrow" (Lambert 2011). As this line of argumentation is more or less the same which is also used by the *Royal Society* in its publication *Climate Change: Evidence and Causes (*The Royal Society 2009), one might get the impression that the inventor cooperates in some form with the *Royal Society*. There are certainly more efficient and less environmentally harmful patents

[2]"the reflecting power of a planet, satellite, or asteroid, expressed as a ration of reflected light to the total amount falling on the surface" (Collins Dictionary 2017).

than this one which meshes deeply into the biosphere of the oceans by inducing environmentally harmful materials. Especially problematic is the introduction of nitrogen and phosphor into the ocean which leads to overfertilization by increasing the rate of algae growth – in addition to the application of chromium, cobalt, nickel, glass, and ceramic as foreign substances. It is highly worrying that such a patent has been approved without a sufficient environmental impact assessment.

Other CDR proposals are aimed at transforming biomass into bio-carbon (or bio-char) through the process of pyrolysis. The resulting solid can be stored on land for a long time or can be used as a biofuel (Uther 2014, with further references).

Further CDR proposals are known by the abbreviations *CCS (carbon capture and storage)* and *BECCS (bioenergy with carbon capture and storage).* CCS is a technical process in which the CO_2 generated during the combustion of fossil fuels is deposited and stored or injected into containers or in liquid form in rock layers. The aim is to emit less CO_2 into the atmosphere, to use fossil fuels in an environmentally friendly manner, and thus to contribute to global climate protection. Although the United States, Norway, and Algeria have already carried out CO_2 storage in emptied petroleum fields or salt-bearing sediment layers for several years, this technology is still in the development stage. The long-term storage possibilities are limited so far, and there are also considerable risks due to the possible escape of the CO_2 from the storage layers. Additionally, the deposition requires a high amount of energy, is very costly, and leads to a partly high reduction in the efficiency of energy generation (CES ifo Group Munich 2017).

At BECCS, the combustion of renewable raw materials is to be combined with CCS technologies (Uther 2014, with further references): "...BECCS is the application of Carbon Capture and Storage (CCS) to any form of bioenergy..." (Smolker and Ernsting 2012). This goes for the production of biomass and coal. CO_2 can be captured by "...chemical and physical absorption, filtering membranes, or adsorption, and may be performed at different stages on a variety of processes..." (Smolker and Ernsting 2012). Promoters of BECCS claim that "...BECCS is one of the few currently available means of removing carbon from the atmosphere...essential for averting climate catastrophe" (Smolker and Ernsting 2012). They consider BECCS "...less risky and more benign..." (Smolker and Ernsting 2012) than "...some proposed climate geoengineering technologies..." (Smolker and Ernsting 2012).

These promoters ignore the obvious disadvantages and risks of BECCS like "... high levels of uncertainty about the possibility of securely storing carbon underground and potential risks to human health and ecosystems..." (Smolker and Ernsting 2012). Furthermore, "...large-scale bioenergy without CCS is already creating an enormous new demand for biomass – a root cause of deforestation, biodiversity loss, human rights violations and escalating carbon emissions..." (Smolker and Ernsting 2012). Additionally, BECCS also leads to "...high additional energy requirements for carbon capture, resulting in significantly more fuel demand to produce the same energy output. Large-scale bioenergy with CCS would thus further increase demand for biomass and worsen impacts..." (Smolker and Ernsting 2012).

Capturing, compressing, and transporting carbon and pumping it underground is very expensive. The storage of CO_2 below ground requires access to underground spaces below land areas and the ocean, facilitating "…a new form of 'underground' land grab…" (Smolker and Ernsting 2012). Some communities have already "…opposed the injection of CO_2 into the ground beneath them…" (Smolker and Ernsting 2012). Furthermore, the difficulty in financing BECCS will lead to dependence on direct government subsidies for a rather long timeframe (Smolker and Ernsting 2012). The lack of reliability of storage and the possible consequences of potential leakages lead to liability and insurance problems. Leakages from flawed capture and emissions from additional energy requirement for injection can cause accidents and natural disasters like earthquakes which can destroy storage formations (Smolker and Ernsting 2012).

Additionally, "…any sudden large release could be extremely dangerous, since exposure to elevated concentrations of CO_2 can be lethal. While normal atmospheric concentration (0.037%) is not toxic, concentrations of 3% and higher result in hearing loss, visual disturbances, labored breathing, headache, impaired vision, and confusion. Concentrations of 20% are quickly fatal, resulting in asphyxiation…" (Smolker and Ernsting 2012). Beyond that, potential leakage of CO_2 from storage formations into overlying freshwater aquifers might occasionally result into a huge increase in dangerous contaminants like arsenic, uranium, barium, etc. (Smolker and Ernsting 2012).

BECCS satisfies "…the oil industry's fast growing demand for cheap continuous supplies of CO_2" (Smolker and Ernsting 2012). The coal industry equally benefits from BECCS by receiving government subsidies for CCS and biomass offering them chances to either stay within their CO_2 allowances in the EU and thus "…avoiding the costs associated with carbon trading…" (Smolker and Ernsting 2012) or protecting themselves from possibly "…stricter and more widely applied future CO_2 caps or pricing mechanisms…" (Smolker and Ernsting 2012).

According to numerous studies, "…large-scale energy including biomass combustion and other processes generally result in even more GHG emissions than the fossil fuels they are intended to replace…" (Smolker and Ernsting 2012). Logging forests and tree plantations for bioenergy and sequestering all the carbon released from harvest, transport, and burning of the previous generation of trees can lead to "carbon debt," encompassing a period of 200 years (Smolker and Ernsting 2012).

The only reasonable and environmentally friendly countermeasure against this "carbon debt" of BECCS would be the reduction of the CO_2 content in the atmosphere by large-scale reforestation and the strengthening of terrestrial ecosystems. This would be a natural alternative to the sometimes recommended but ecologically highly problematic "reforestation" with nonlocal and genetically manipulated plants, an artificial countermeasure which pertains to CDR measures because industrial tree plantations for bioenergy as a kind of "green deserts" require synthetic fertilizers and other agrochemicals, depleting freshwater and soils, and thus cannot be compared to naturally grown forests (Smolker and Ernsting 2012; Uther 2014, with further references).

CE schemes typically strive for the reduction of Climate Change impacts by using "analogues of processes already present in the Earth system" (Thornes and Pope 2014). Stratospheric particle injection as one kind of SRM technology, for example, "...mimics the effects of large volcanic eruptions; these eruptions can inject mega-tonnes of particle material into the stratosphere leading to an increased global albedo and planetary cooling..." (Thornes and Pope 2014). The last big volcanic eruption on the Earth was the one of Mt. Pinatubo (Philippines) in 1991, reducing mean global surface temperature by some 0.5 °C for a period of 1–2 years. The CE analogue to volcanic eruptions is implemented by anthropogenic particle injections into the stratosphere via non-volcanic route such as an aeroplane or pipe delivery system (Thornes and Pope, with further references; Uther 2014, with further references).

Following the observation of the effects of volcanic eruptions on the climate, the renowned atmospheric scientist *Paul J. Crutzen* published an editorial essay titled *Albedo Enhancement by Stratospheric Sulfur Injections. A Contribution to Resolve A Policy Dilemma*. In this essay, Crutzen argues as follows: "...If sizeable reductions in greenhouse gas emissions will not happen and temperatures rise rapidly, then climate engineering, such as presented here, is the only option available to rapidly reduce temperature rises and counteract other climatic effects..." (Crutzen 2006).

This controversial essay by *Paul Crutzen* sparked a lively discussion in scientific circles, causing a significant increase in scientific CE-related publications from 2006 to 2010 in international journals and science magazines (Uther 2014, with further references). Additionally, Paul Crutzen's essay was taken up by the media with irony and skepticism. The Guardian titled: "The atmospheric scientist Paul Crutzen would like to save the world and darken your day" (Uther 2014, with further references).

CE remains controversial because of its chances for countermeasures against Climate Change on the one hand and its risks and side effects on the other hand. For this reason, it is important that Climate Change scientists monitor further advances in CE research. Many of the latest CE trends can be found on the "Climate Engineering News" website of a renowned German research institute (Kieler Earth Institute 2017).

Beyond that, it is striking that the British print media report on CE often by using metaphorical expressions borrowed from the military vocabulary, e.g., "Miracle weapon against Climate Change," "particle bombs," or "sulfur bombs in the stratosphere" for SRM technologies (Uther 2014, with further references). This can also be observed in the German media which, among other things, write about "Techno-Creationists on the Climate Front" or "Star War" against Climate Change in an allusion for weapons in space against which also the *Brundtland Report* had warned (Uther 2014, with further references; Schwarz-Herion 2015, with further references).

This choice of words, the intensive involvement of the military- above all the British *Royal Air Force (RAF)* - into CE research, and the fact that the spraying of sulfate aerosols is especially eagerly discussed in military circles (Schwarz-Herion 2015, with further references) should make an attentive and critical observer suspicious and suggest that there might exist covert field trials for the use of weather weapons or even covert warfare. The next section of this chapter will deal with this rather neglected aspect.

12.4 Covert Warfare by Weather Modification and Climate Engineering: "Owning the Weather by 2025" – Science Fiction Fantasy or Fact?

As shown in the introduction, natural disasters and extreme weather patterns have visibly accumulated over the last decades, especially in the last 20 years: Additionally to the disasters mentioned above, there had been the tornado disaster in Texas from July 1997 causing 29 deaths and an estimated $20 million in personal and commercial insured losses (CDC 1997), furthermore *Hurricane Georges* traveling across the Caribbean before striking in the Dominican Republic and Haiti on September 22, 1998 (US Agency for International Development 1998), and the earthquake in Chile in 2010 affecting two million people including over 300 persons who got killed (CNN 2010). Furthermore, there were severe floods in Texas in 2015 (The Millennium Report 2015; Duclos 2015; Hauser 2016) and dramatic floods in Europe in 2016 (McElwee 2016; Euronews 2016) which are only some of the better known natural disasters in recent times. Furthermore, according to the global reinsurance firm *Munich Re,* the year 2017 has already seen 49 separate weather, climate, and flood disasters (Rice 2017). The 2017 weather disasters also included gigantic *X-ray tsunami* rolling through galaxy cluster (Wall 2017) and a major winter storm moving across the continental United States (Spaleta 2017).

Although the IPCC, the Royal Society, and other groups of influential scientists are rather convinced that the "anthropogenic[3]" greenhouse effect reportedly caused by CO_2 emissions from burning fossil fuels is most probably the main reason for these severe natural disasters and the extreme weather patterns (NOAA 2015; IPCC 2014; see also above Sect. 12.2), there might also be alternative explanations for a different kind of anthropogenic greenhouse effect. Might it also be possible that some of the more recent natural disasters might have been caused or at least enhanced by the targeted covert use of already existing weather modification technology in the framework of field research or even in a targeted way by covert weather warfare? In other words: Might the ***mono-causal*** global-warming-by-GHG hypothesis due to regular human activities, e.g., GHG emissions from burning fossil fuels for industry, traffic, and households, serve as a distraction from deliberate weather modification activities for the aforementioned reasons?

At first glance, one might think that the idea to use targeted weather modification as a weapon of war would be nothing but an utopia, laid down in literature such as *Scientific Outlook* written by *Bertrand Arthur Russell* in 1931, in the 1952 book *The Next Million Years* by *Charles Darwin*'s grandson *Charles Galton Darwin* where he talked about gaining direct control over the climate (Rich 2011), or in science fiction movies like Emmerich's *Noah's Ark* from 1984 where the CIA controls the

[3] Strictly speaking, the term "anthropogenic greenhouse effect" is misleading because it implies that humans would *cause* the greenhouse effect in the first place, whereas the greenhouse effect is a *natural phenomenon*. More appropriate and precise would be the term "anthropogenic enhancement of the natural greenhouse effect."

weather (see Chap. 1, Sect. 1.5). In any case, early patents for weather modification technology have already existed around the turn of the nineteenth century (see above Sect. 12.3.2).

Actually, attempts to modify the weather in real life exist since the late 1940s at least (Rich 2011). Since then, a range of projects involving acts of weather modification for field experiments or even for actual use at war have been carried out (Rich 2011). Declassified records reveal that – in the period from 1949 to 1955 – the *Royal Air Force (RAF)* released various substances including dry ice (=solid CO_2), silver iodide, and salt into the atmosphere at high altitudes in order to induce rain (Coles 2010).

The military race during the *Cold War* fostered all kinds of innovations by scientists. Apart from nuclear physics, e.g., bombs and submarine reactors, as well as solid-state physics, e.g., electronics and metallurgy, priority in funding was also given to physical geoscience; military officers were interested in understanding "...almost everything about the environments in which they operated, from the ocean depths to the top of the atmosphere...operational data and basic research results were often the same thing" (Weart 2017).

Especially the US government fostered all kinds of geophysical work with a focus on meteorology because weather has played a major role in warfare since antiquity: "...After the war, military agencies joined civilian ones in fostering research that might eventually improve weather prediction. The work ranged from better data-collecting networks to laboratory studies of radiation and attempts to model weather on digital computers..." (Weart 2017).

Additionally to the daily forecast, some experts envisioned plans to change the weather deliberately, e.g., by helping farmers to "seed" clouds with silver iodide smoke, in hope of making rain or harming an enemy by droughts or enduring snows. Around 1955, some scientists, above all the mathematician and nuclear bomb expert *John von Neumann*, issued warnings that "...'climatological warfare' could become more potent than nuclear war itself..." (Weart 2017, with further references).

In *Jacob Darwin Hamblin*'s book *Arming Mother Nature* which won the *2014 Paul Birdsall Prize from the American Historical Association for best book in military or strategic history*, the author, director of the Graduate Program in Environmental Arts and Humanities at *Oregon State University*, reveals that "... military planning for World War III essentially created 'catastrophic environmentalism': the idea that human activity might cause global natural disasters..." (Hamblin 2013). Hamblin states that this idea "...emerged out of dark ambitions, as governments poured funds into environmental science after World War II, searching for ways to harness natural processes to kill millions of people..." (Hamblin 2013). The objective of weaponizing the weather included the creation of artificial tsunamis and the **targeted** melting of the ice caps to "drown coastal cities" (Hamblin 2013), furthermore putting huge areas of vegetation into fire and changing local climates in a targeted way by nuclear weapons, as well as Oxford botanists advising British generals about possible ways to destroy enemy crops during the war in Malaya (Hamblin 2013).

Some early examples of severe human made weather disasters in real life in the first half of the twentieth century were the following projects:

- *Project Cirrus*: This project of the *US Army*, *Navy*, and *British Royal Air Force* under the leadership of *General Electric (GE)*, conducted in the late 1940s, caused snow artificially by dropping 1.4 kg of dry ice pellets (=solid CO_2) from an airplane into a super-cooled stratus cloud and modified a hurricane by introducing 80 lbs. of dry ice into the hurricane so that it changed direction and began traveling back toward the east coast of the United States, hitting Savannah (GA) with recorded gusts of 85 MPH, flooded coastal areas, caused millions in damage, and killed at least one person (The Black Vault 2015; Hap 2013a).
- *Project Cumulus* (1949–1955): According to declassified records, the *Royal Air Force (RAF)* released various substances including dry ice, silver iodide, and salt into the atmosphere at high altitudes in order to induce rain. These cloud-seeding experiments over southern England were reportedly responsible for the 1952 flood in the Devon village of Lynmouth resulting in 34 deaths and the destruction of diverse structures (Rich 2011; Coles 2010).

Interestingly, soon after – in the late 1950s – some scientists started to push the *global-warming-by-CO_2 hypothesis* in public settings, e.g., Edward Teller who "… told an assembly of scientists in 1957 that rising CO_2 levels might eventually melt back the polar ice caps and inundate the world's lowlands…" (Weart 2017; Hamblin 2013), clearly having "a personal stake as a nuclear expert" (Weart 2017). Apparently, there have been different stakeholders and different motives to create incentives to promote this hypothesis since the 1950s already: (1) nuclear lobbyists in order to distract from the risks of nuclear energy by stressing or even over-exaggerating **only** the disadvantages of fossil fuels and (2) persons and entities involved into weather modification experiments for military applications with the objective to conceal these experiments (Weart 2017; Hamblin 2013).

The 1960s have seen further progress in innovative weather modification technology which had the potential to be used as a weapon of war and which would actually be used as such in the late 1960s. A patent targeted at the deliberate artificial enhancement of the quantity of CO_2 in the atmosphere, *Patent US 2963975*, "...a cloud seeding bullet of the type wherein liquid carbon dioxide under pressure is released…after it leaves the gun from which it is fired…" (Musser 1960), was granted to *C. Walton Musser*, assignor to the United States of America as represented by the *Secretary of the Army* in December 1960 (Musser 1960).

In early 1962, the acknowledged meteorologist *Harry Wexler*, a Harvard and MIT graduate in meteorology and then *Chief of Scientific Services* of the *US Weather Bureau*, mentioned at the beginning of a speech entitled "On the Possibilities of Climate Control" that "…the subject of weather and climate control is now becoming respectable to talk about" (Fleming 2007, with further references). Wexler also cited some extracts from President John F. Kennedy's speech at the UN in which Kennedy proposed "…cooperative efforts between all nations in weather prediction and eventually weather control…" and the State Department's urgent demand to do an "…early and comprehensive study in the light of developments in outer space of the possibility of large-scale weather modification…" (Fleming 2007, with further references). In his speech, Wexler pointed out that cloud modification technology facilitating mostly localized modification of precipitation was acceptable but raised serious concern about a global "…manipulation of the Earth's shortwave and long-wave radiation budget that would result in 'rather large-scale effects on general circulation patterns in short or longer periods, even approaching that of climatic change'…" (Fleming 2007, with further references).

Such "large-scale effects" also involved raising and lowering world temperature by several degrees as well as the destruction of all ozone in the stratosphere above the arctic circle using a rather small amount of catalytic agents, e.g., chlorine or bromine. Wexler was equally interested in unintentional climate modification, for instance, by rocket exhaust gases or other kinds of pollution, and the targeted peaceful and hostile effects of weather and climate modification (Fleming 2007, with further references).

Wexler who was also chief scientist of the *US Antarctic program for the International Geophysical Year* founded the *Mauna Loa Observatory*, helped *Dave Keeling's* measurements of CO_2, and wrote on peaceful use of satellites and weather control for John F. Kennedy. Wexler's plans for the *World Weather Watch* would only be implemented in 1963, one year after his death (Fleming 2007, with further references).

In his speeches about *climate control* in 1962, Wexler addressed new scientific developments including computing and satellites, convincing him that the manipulation and control of large-scale phenomena in the atmosphere were excellent possibilities, e.g., the increase in the amount of CO_2 emissions, referring to a study from 1961 about the use of rockets to modify the Earth's upper atmosphere (Fleming 2007, with further references).

In spite of recognizing these possibilities, Wexler realized "a growing anxiety" in public statements that "…Man, in applying his growing energies and facilities in the power of the winds and storms, may do so with more enthusiasm than knowledge and so cause more harm than good…" (Fleming 2007, with further references). The scientist selected only some possibilities "…limited primarily to interference with the Earth's radiation balance on a rather large scale …", discussing "…in a purely theoretical framework those atmospheric influences that man might attempt deliberately to exert and also those which he may now be performing, perhaps in ignorance of its consequences…" (Fleming 2007, with further references). Wexler pointed out: "…We are in weather control now whether we know it or not…" (Fleming 2007, with further references).

Furthermore, Wexler discussed technologies to manipulate the atmosphere's radiation heat budget, covering, among others, the following possibilities:

- Raising global temperature by 1.7 °C via injecting a cloud of ice crystals into the polar atmosphere by detonating 10 H-bombs in the Arctic Ocean as described in an article in *Science* magazine (Wexler 1958)
- Decreasing global temperature by 1.2 °C by introducing a ring of dust particles into the equatorial orbit to darken the Earth, a modification of an earlier Russian

proposal to increase the temperature in the Artic (Fleming 2007, with further references)

- The depletion of all stratospheric ozone, moving the tropopause[4] upward and "... cooling the stratosphere by up to 80 °C by an injection of a catalytic agent such as chlorine or bromine..." (Fleming 2007, with further references)

Wexler was particularly worried about an unintentional damage to the ozone layer by growing rocket exhaust polluting the stratosphere and about future near-space experiments getting out of hand, taking reference to *Operation Argus*, nuclear blasts in near space (1958); *Project West Ford*, a ring of copper dipole antennas in orbit (1961); as well as *Project Highwater*, ice crystals injected into the ionosphere (1962), as examples for interventions with unforeseeable risks (Fleming 2007, with further references). He also saw the possibility of deliberately causing damage or even *geophysical warfare* by attacking the ozone layer over a rival nation. Wexler ended his speech with the explanation that he was not doing proposals for intervention but only studied the basic equations and engineering aspects of general circulation research like the natural behavior of the atmosphere, inadvertent effects, and those issues which were of special interest to the *Department of Defense (DoD)* (Fleming 2007, with further references).

At the end of his speech, Wexler issued the following warning:

> ...Climate Control can best be classified as 'interesting hypothetical exercises' until the consequences of tampering with large scale atmospheric events can be assessed in advance. Most such schemes that have been advanced would require colossal engineering feats and contain the inherent risk of irremediable harm to our planet or side effects counterbalancing the possible short-term benefits. (Fleming 2007)

Wexler died prematurely in 1962 at the age of 51, but his knowledge lives on in the *Library of Congress* and impresses also renowned modern ozone scientists like Nobel Laureates *Sherwood Rowland*, *Paul Crutzen*, and *National Academy of Sciences* President (NAS) *Ralph Cicerone* (Fleming 2007, with further references).

You see from the items mentioned above that Wexler was a visionary pioneer of CE who had been intelligent and enlightened enough to see the immense dangers and the intrinsic risks of this innovative technology. Therefore, it is difficult to understand why Wexler's name is so rarely mentioned in connection with weather modification and CE technology. Apparently, Wexler's involvement into military projects is supposed to be concealed from the public eye.

In 1966, a recommended national program in weather modification was provided by a NASA member in the framework of a report to the *Interdepartmental Committee for Atmospheric Sciences* (NASA 2013 with reference to Newell 1966) – allegedly to develop countermeasures against the destructive effects of weather modification technology and to enhance its beneficial effects. Nevertheless, only one year later, *Operation Popeye* started – a then top secret campaign of weaponized weather modification – which would be conducted from 1967 to 1972 to extend the monsoon season over North Vietnamese and Viet Cong resupply routes throughout southeast

[4] "the plane of discontinuity between the troposphere and stratosphere" (Collins English Dictionary 2017).

Asia, particularly over the supply route "Ho Chi Minh Trail" by seeding units composed of silver iodide developed at the *Naval Weapons Center* in *China Lake California* with the objective to trigger a huge amount of mud over supply routes (Hap 2013b; Rich 2011).

Having the potential of further advances in weather modification technology in mind, Zbigniew Brzezinski predicted in 1970 in his book *Between Two Ages*:

> ...Technology will make available, to the leaders of major nations, techniques for conducting secret warfare, of which only a bare minimum of the security forces need be appraised ... Techniques of weather modification could be employed to produce prolonged periods of drought or storm.... (Brzezinski 1970)

Furthermore, in the early 1970s, an especially powerful weather modification patent was developed to change the climate in certain regions of the Earth significantly: *United States Patent 3,564,253 – SYSTEM AND METHOD FOR IRRADIATION OF PLANET SURFACE AREAS*,[5] invented by *Arthur G Buckingham*, published on February 16, 1971, was assigned to *Westinghouse Electric Corporation*, Pittsburgh, PA. This patent is described as follows (extract):

> ...system and method for generalized irradiation of relatively large surface areas of a planet, such as the earth, the moon, etc., for illumination, heating, weather control, etc., employing one or more planet-orbiting self-erecting planar-reflector satellites controlled in attitude and orbit position to reflect energy from the sun to a desired area on the planet's surface...a satellite means launcheable into orbit around such planet...including a compacted membranous solar-energy reflector means unfurlable in space to form a rigidized planar reflector surface, and attitude-changing means for orientating said plane surfaced reflector to intercept solar energy from the sun and reflect same onto a selected area of said planet for a selected period of time during each orbit.... (Buckingham 1971)

The field of the invention concerns the "radiation of energy in the form of electromagnetic waves for heating and/or illumination." The patent enables the modified reflection to be maintained "...for prolonged or controlled periods during nighttime and/or daytime" (Buckingham 1971). Especially revealing is the following statement done by the inventor:

> ...As to heating and weather control, it is well known that a major causative factor in the generation of weather patterns and climate is solar radiation...if the amount of solar radiation which is received in a given area is varied significantly by satellite reflector means, some variation in the temperature and/or weather patterns will result. A crude estimate of the number of satellites with 100 -ft. diameter planar-reflectors 20 required in a synchronous orbit to produce a temperature rise of 1.3° R. (= 1.625 °C) in a circular area with a diameter of 208 miles is approximately 12,000 if there is no appreciable mass flow in and out of the area. This radiation intensity represents about one percent of the solar radiation.... (Buckingham 1971)

So, the inventor of this patent strived for a temperature rise of 1,625 °C. Interestingly, this increase in temperature would even exceed the increase in temperature which – according to the IPCC, the Royal Society, and some other experts –

[5] This patent had the potential to use a technology which would later on be known as solar radiation management, SRM.

might allegedly result from the inadvertent "anthropogenic greenhouse effect" (IPCC 2014; The Royal Society 2014; see also above and Chap. 1).

Equally in 1971, *The Washington Post* published a story about *US Air Force* "rainmakers" in Southeast Asia. This story was later on confirmed by the publication of the *Pentagon Papers* in the *New York Times*, mentioning weather warfare by project *Pop Eye* in the Vietnam war. This publication came out soon after the top secret *Pentagon Papers* had been leaked by the economist and RAND analyst *Daniel Ellsberg* who had worked on these papers, originally titled *US Decision-Making in Vietnam, 1945–1968* which had been ordered by the then *Defense Secretary Robert McNamara* (Ellsberg 2017; Hap 2013b).

Consequently, the *US Senate* called for an international agreement "…prohibiting the use of any environmental or geophysical modification activity as a weapon of war…" (UNOG 1976). Soon after, in the *International Convention on the Prohibition of Military or any Other Hostile Use of Environmental Modification Techniques (ENMOD Convention)*, the UN banned "...the military or otherwise hostile use of weather modification techniques having widespread, long-lasting or severe effects…"(UNOG 1976, 1978). Among weather phenomena which could be created by environmental modification techniques already back then, the *ENMOD Convention* listed, inter alia, earthquakes and tsunamis, changes in weather patterns (clouds, precipitation, cyclones, and tornadic storms), changes in climatic patterns, changes in the state of the ozone layer, and changes in the state of the ionosphere, stressing that this list was ***not*** conclusive (UNOG 1976). The *ENMOD Convention* was approved by the *General Assembly of the United Nations* in 1976 and entered into force in 1978 (UNOG 1978).

Nevertheless, in spite of the *ENMOD Convention*, military research into weather modification advanced even further in the 1980s: The *Central Intelligence Agency (CIA)* did research on the possible effects of global warming on Soviet harvests, and *Cold War* scientists learned to think globally and became aware of humanity's possibility to change the natural environment (Hamblin 2013). A large number of patents for weather modification techniques suitable for military application were granted by the *US Patent and Trademark Office (USPTD)*. Among the most interesting patents suitable for military weather modification granted by the USPTD is *United States Patent 4,686,605 Eastlund* – "Method and apparatus for altering a region in the earth's atmosphere, ionosphere, and/or magnetosphere." This patent was granted to *Advanced Power Technologies, Inc. (APTI)* on August 11, 1987 (Eastlund 1987), over two decades after nuclear physicist *Edward Teller* had already boasted: "We know how we can modify the ionosphere…We have already done it" (Hamblin 2013).

One year after the *Eastlund Patent* was granted, *The New York Times* published the article *Global Warming Has Begun, Expert Tells Senate* (Shabecoff 1988; see also Chap. 1, Sect. 1.3) where *James E. Hansen* from *NASA* claimed that global warming was due to man-made GHG emissions (Shabecoff 1988; see also Chap. 1, Sect. 1.3). Later on, some individuals and institutions which had once helped the *Pentagon* in the development of weather modification technology like *Eduard Teller* would propose technological solutions to cope with "severe failures of the climate"

(Teller et al. 1997) like global warming by GHG or the next Ice Age (Teller et al. 1997; Hamblin 2013, with further references).

In the 1990s, even more sophisticated environmental modification technologies for military use were developed (Rich 2011). This included, above all, the *High Frequency Active Auroral Research Project (HAARP)*. After HAARP[6] had been approved by the *US Congress* for funding in 1990 by its defense spending bills, research for HAARP began in 1992 by *Advanced Power Technologies, Inc. (APTI)* as holder of the aforementioned *Eastlund Patent*, before being sold to *E-Systems* and then to *Raytheon Corporation* along with the *Eastlund Patent* (Rich 2011). HAARP was designed and built by *British Aerospace Systems (BAES)* and *Raytheon Corporation*[7] (Rich 2011).

The HAARP facility was originally located in Gakona, Alaska, and consisted of "...a megawatt radio transmitter and worked by heating the ionosphere, an active, electrically-charged part of the upper atmosphere 40 to 500 miles above the earth which acts as a shield against the bombardment of high-energy particles from space...one of the most powerful multi-purpose weapons developed by the US military...acknowledged as a 'weather weapon' by parts of the mass media..."(Rich 2011, with further references). It was managed by the *US Air Force Research Laboratory (AFRL)*, the *Office of Naval Research (ONR)* of the *US Navy*, the *Defense Advanced Research Projects Agency* (DARPA), and the *University of Alaska's Geophysical Institute*. The management team consisted of scientists from universities, the private sector, and government (Rich 2011).

The 1990s have also seen a range of other research projects in weather modification and CE technology. Only a representative selection of these projects is listed in the following:

- In 1994, a *US Air Force JNLWD Report* described weather warfare as a possible weapon of mass destruction and the *US Navy's* 1994 *Nonlethal Warfare Proposal* suggested the artificial creation of floods, blizzards, and droughts to destroy a nation's economy (Rich 2011).
- In August 1996, the *Defense School* research paper *Weather as a Force Multiplier. Owning the Weather in 2025* was presented to the *US Air Force*. Although this paper focuses primarily on localized weather modification and its incorporation in war-fighting capability, it also envisions the widest range of weather modification opportunities for the United States' military including the "...alteration of global climate on a far-reaching and/or long-lasting scale..." (House et al. 1996).

[6] Based on Eastlund's patents and thus indirectly on discoveries made by Dr. Nikola Tesla, the famous inventor and winner of the 1915 Nobel Physics Prize whose findings included wireless technology and alternating current electricity.

[7] Some wonder if Raytheon's high-level employees, whose names appear on the passenger lists of American Airlines Flights 11 and 77 and United Airlines Flight 175 on 9/11 2001, might have been bearers of secret or even patent holders of HAARP patents: On Flight 11, there had been Raytheon's senior mechanical engineer, its vice president of operations electronic systems, and its senior quality control engineer; on Flight 77 Raytheon's director of program management; on United Airlines Flight 175 a Raytheon Corporate Executive (*USA Today* 2001).

- In October 1996, the renowned scientific journal *Nature* published an article, mentioning Russia's "ambitious research into 'tectonic warfare'…involving attempts to stimulate 'artificial' earthquakes as weapons of destruction…" (Levitin 1996).

The then *Secretary of Defense William Cohen* probably had these projects in mind when he warned against terrorists engaging in an "…eco-type of terrorism whereby they can alter the climate, set off earthquakes, volcanoes remotely through the use of electromagnetic waves…wreak terror upon other nations…" in a public discussion at the *Conference on Terrorism, Weapons of Mass Destruction, and US Strategy* at the University of Georgia on April 28, 1997 (Cohen 1997).

The HAARP project (see above) was conducted simultaneously to the aforementioned projects throughout the 1990s. It continued until 2013, when the plant in Alaska was closed down and their tasks were basically taken over by DARPA, the *Defense Advanced Research Projects Agency* (DARPA 2017; Schwarz-Herion 2015, with further references).

Equally in 2013, one year after the *U.S. Army* had developed a powerful *lightning laser weapon* thouroughly tested in January 2012, capable of hitting targets which conduct electricity better than the air or ground surrounding them, using 50 billion Watts of optical power and forming an electro-magnetic field to create plasma (BBC 2012), *The Independent* reported about a joint project of the *Central Intelligence Agency (CIA)* and the *National Academy of Sciences (NAS)*. Allegedly, the objective of this project was to investigate if humans could use CE to stop Climate Change (Williams 2013).

Soon after the launch of this project, a series of weather disasters started in Asia, the United States, and Europe. These weather disasters included 32 lightning deaths in India during a storm in October 2013, 23 lightning deaths in the U.S. over the course of 2013 (Quinn 2013), and a severe lightning strike in Northern Germany in May 2013 injuring 39 out of 500 persons gathered for an Open Air *Father's Day* Party (Agence France Presse 2013), the *Kelantan flood* in Malaysia (see Chap. 2; Davies 2015), the *Texas floods* of 2015 (The Millennium Report 2015; Duclos 2015) and 2016, further floods simultaneously happening in the United States, e.g., in Texas (Hauser 2016), in several European countries, above all in Germany and France, requiring the rescue of hundreds of people including children and even killing several people (McElwee 2016; Euronews 2016): In Paris, the River *Seine* rose above 5 m, forcing rail operator *SNCF* to close a metro line down used by tourists to get to the *Eiffel Tower* and *Notre Dame Cathedral* and requiring the temporary closure of the *Louvre Museum* to bring irreplaceable artworks into safety (Euronews 2016). Equally in 2016, lightnings in Germany and France hurt almost 50 persons – especially children – in Southern Germany and Paris on a single weekend; 35 of them had to be hospitalized (*Die Welt Online* 2016). Severe weather disasters in China, Pakistan, and India killed 130 persons and destroyed thousands of buildings and hundreds of thousands of hectare of agricultural land in 2016 (*BNN* 2016).

Might these natural disasters taking place simultaneously in different countries and continents have partially been the results of successful field trials in the framework of the aforementioned NAS/CIA project, consisting of simultaneous application of weather modification techniques on different spots of the globe by mature weather modification technology as the ones envisioned for 2025 by the authors of *Owning the Weather in 2025*? In any case, the technology for weather modification and CE is there, it has been tested in practice, and the will of influential entities to optimize this technology to alter the climate and the weather on a global base for military and other hostile purposes (like reduction of the world population, see Chap. 1, Sect. 1.6) cannot be ignored either.

Already one year before the joint CE-project of NAS/CIA, the *Planungsamt der Bundeswehr* (German Army) had discussed the political implications of CE in its study *Future Topic Geo-Engineering* (2012). Based on the intrinsic risks of CE, this study encouraged a binding international legal regulation of CE (Planungsamt der Bundeswehr 2012).

In 2016, the Cypriot government accused the British *Royal Air Force (RAF)* of "...interfering with the weather to enable its *Tornado* and *Typhoon* aircraft to fly missions to Syria and Iraq in clear conditions..." (Daily Mail 2016). Actually, local weathermen had recently forecasted heavy rain only for it to remain dry. British forces have been accused of cloud-seeding, i.e., of using a chemical to divert rain clouds, and agriculture minister *Nicos Kouyialis* ordered an official inquiry, telling a Cypriot parliamentary committee: "...Such actions could change the climate of the region as they may change the earth's atmosphere..." (Daily Mail 2016).

Some months later, in June 2016, *CIA Director Brennan* spoke about CE at the *Council on Foreign Relations (CFR)*: "...On the geopolitical side, the technology's potential to alter weather patterns and benefit certain regions at the expense of others could trigger sharp opposition by some nations..." (Brennan 2016). A certain challenge in this respect, however, is the fact that "...it is widely taken for granted that the military sector is beyond the reach of a state's civil sector, both in democratic and totalitarian states..." (Westing 2013).

There are strong indicators that those who covertly use weather modification and CE technology for military or other hostile purposes tend to deny the efficiency of this technology. In 2001, the British *Ministry of Defence (MoD)* had still played down the efficiency of cloud seeding by alleging: "...Cloud seeding has rarely been successful anywhere in the world. Consequently, the *Met Office* had not pursued this line of research for many years..." (Coles 2010, with further references).

In 2010, the *British Ministry of Defence (MoD)* contradicted itself by doing the following confession in a public document, announcing that by 2040:

> ...Weather modification will continue to be explored...Manipulation of the weather may affect changes in operating conditions, limit aviation flight envelopes, generate poor visibility while providing concealment and disrupt line of communications. Weather modification may also affect morale...The WMO Atmospheric Research and Environment Programme notes that there are several operational programmes in fog dispersion, rain and snow enhancement, as well as hail suppression.... (Coles 2010, with further references)

Critics claim that – together with the ambitious plans of the Pentagon – "…to achieve Full Spectrum Dominance by 2020…" (Coles 2010), the fact that the British MoD and the US Air Force had the capability to destroy an entire village with 90 million tonnes of rainfall as far back as the 1950s, indicates that "…weather weaponisation is being used under the cover of anthropogenic global warming…" (Coles 2010).

It might also be interesting to note that Patent *US 20030085296 A1-Hurricane and Tornado control device,* involving audio generators capable of disrupting, enhancing, or directing the formation of a low atmospheric weather system, was registered in 2001 and published in May 2003 (Waxmanski 2003), i.e., ca. 2 years before the *Hurricane Katrina* disaster. This patent was followed by numerous other storm modification patents in the years 2007 to 2015 (IFI Claims Patent Services 2017).

Political Leaders, diplomats, and scientists from nations who have been affected by ecotype terrorism or covert ecological warfare or who fear such attacks in the future should seek the dialogue with colleagues from other nations who were equally attacked or fear future attacks. They might discuss coordinated actions to seek a constructive dialogue and develop technological, political, economic, diplomatic, and military counter-strategies to defend the people and nature against this kind of attacks in the future. In this connection, a modification of the ENMOD Convention by prohibiting all kinds of CE and weather modification techniques involving incalculable risks, regardless of the intention of their application (hostile or peaceful), might become necessary to tackle this problem successfully on a global base.

12.5 Conclusion

The numerous risks of CE outweigh the chances for climate mitigation or the prevention of Climate Change by far. Therefore, these high-tech solutions should be carefully designed, planned, and implemented and only be used, if any, in a very restrictive way. Apart from ecological and economic risks, political aspects should also be taken into consideration because weather modification and CE technologies can be useful for some nations and regions but damaging for others, even opening the door to covert weather warfare.

This problem could only be overcome by binding international regulations, eventually forbidding or responsibly regulating the use of such weather modification and CE technology tools which can be abused for covert warfare or even acts of international terrorism, even if originally designed and intended for peaceful purposes. In sum, it is better to resort to conventional means like legal regulations for the reduction of GHG and enhancing sinks by rigorous reduction of deforestation and implementing large-scale reforestation with naturally growing site-specific plants on a global base, but – in contradistinction to the controversial *Wildland Project* (see Chap. 1, Sect. 1.3) – in an economically reasonable and socially acceptable way by giving preference to uninhabited areas and to unused areas owned by large landowners who sell parts of it voluntarily for the general benefit, thus avoiding conflicts with property rights and farmland while allowing humans to live in harmony with nature.

References

Agence France Presse (2013) Lightning strike hurts 39 at Father's Day party in Germany. In: Captital FM Kenya, May 9, 2013. Retrieved on July 14, 2017, from https://www.capitalfm.co.ke/news/2013/05/lightning-strike-hurts-39-at-fathers-day-party-in-germany/

BBB Better Business Bureau (2017) This business is not BBB accredited. Dyn-O-Mat, Inc. Retrieved on July 10, 2017, from https://www.bbb.org/south-east-florida/businessreviews/auto-services/dyn-o-mat-in-jupiter-fl-20001515

BBC (2012) Lightning laser weapon developed by US Army. On: BBC-Technology, June 28, 2012. Retrieved on July 14, 2017, from http://static.bbci.co.uk/news/1.223.02234

Belitz H-D (1999) Food chemistry. Springer, Berlin/Heidelberg

Biello (2009) Pulling CO2 from the Air: Promising Idea, Big Price Tag. Yale School of Forestry and Environmental Studies, New Haven

Blake ES, Kimberlin BT, Berg RJ, Cangialosi JP, Beven JL II (2013) Tropical cyclone report hurricane sandy (AL 182012). National Hurricane Center, Miami

BNN (2016) Unwetter wüten in China und Indien. BNN July 4, 2016, edition 152, p. 7

Brennan JO (2016) Director Brennan speaks at the council on foreign relations. Remarks as prepared for delivery by central intelligence agency director John O. Brennan at the Council on Foreign Relations, Washington, DC. Retrieved from https://www.cia.gov/news-information/speeches-testimony/2016-speeches-testimony/director-brennan-speaks-at-the-council-on-foreign-relations.html on July 26, 2017

Brzezinski Z (1970) Between two ages. America's role in the technetronic era. The. Viking Press, New York

Buckingham AG (1971) United States patent 3564253 – SYSTEM AND METHOD FOR IRRADIATION OF PLANET SURFACE AREAS. Retrieved from http://www.freepatentsonline.com/3564253.html on July 22, 1971

Caldeira K (2007) Climate engineering is doable, as long as we never stop. In: Smalley E (ed.) Wired. Retrieved from https://www.wired.com/2007/07/geoengineering-7/ on June 5, 2017

CDC (1997) Tornado Disaster – Texas, May 1997. Retrieved from https://www.cdc.gov/mmwr/preview/mmwrhtml/00049839.htm on July 20, 2017

CES ifo Group Munich (2017) Carbon Capture and Storage (CCS) Retrieved from https://www.cesifo-group.de/de/ifoHome/facts/Glossar/08-Umwelt-Klima-Energie/Carbon-Capture-and-Storage-CCS.html. on July 23, 2017

CIA (2002) Intelligence update: chemical warfare agent issues. https://www.cia.gov/library/reports/general-reports-1/gulfwar/cwagents#. Accessed on 27 July, 2017

CNN (2010) More than 2 million affected by earthquake, Chile's president says. Retrieved from http://edition.cnn.com/2010/WORLD/americas/02/27/chile.quake/index.html on July 20, 2017

Cohen W (1997) WS News transcript. Press operations. DoD News briefing: secretary of defense William S. Cohen April 28, 1997 8:45 AM EDT. Presenter: decretary of defense William S. Cohen April 28, 1997 8:45 AM EDT. Retrieved on July 26, 2017, from http://archive.defense.gov/Transcripts/Transcript.aspx?TranscriptID=674

Coles TJ (2010) Weather weapons: the dark world of environmental warfare. Retrieved from AntiCorruption Society on June 14, 2016: https://anticorruptionsociety.com/tag/climate-change/

Collins English Dictionary (2017) Harper Collins Publisher, New York

Cordray JM (1914) United States patent 1103490 "RAIN-MAKER". Inventor and assignee: JAMES M. CORDRAY. Retrieved from http://www.freepatentsonline.com/1103490.html on June 4, 2017

Crutzen P (2006) Albedo enhancement by stratospheric sulfur injections: a contribution to resolve a policy dilemma? Climatic Change 77(3):211–219

Daily Mail (2016) RAF stole our rain, says Cyprus as British military bizarrely accused of interfering with the weather so Tornado and Typhoon aircraft can fly. In: Daily Mail, published on February 21, 2016. Retrieved from http://www.dailymail.co.uk/news/article-3456522/RAF-stole-rain-says-Cyprus-British-military-bizarrely-accused-interfering-weather-Tornado-Typhoon-aircraft-fly.html#ixzz4j7OuRwfI. on July 21, 2017

DARPA (2017) Creating breakthrough technologies for national security. https://www.darpa.mil/. Accessed on 25 July, 2017

Davies R (2015) Malaysia floods – Kelantan flooding worst recorded as costs rise to RM1 billion. Published on 5th of January, 2015. http://floodlist.com/asia/malaysia-floods-kelantan-worst-recorded-costs. Accessed on 28 July, 2017

Der Spiegel online (2009) Algendüngung weniger wirksam als gedacht. Retrieved from http://www.spiegel.de/wissenschaft/natur/co2-speicherung-im-ozean-algenduengung-weniger-wirksam-als-gedacht-a-604235.html. on July 25, 2017

Die Welt (2016) Viele Verletzte nach Blitzeinschlägen in Deutschland und Frankreich. http://www.welt.de/newsticker/dpa_nt/infoline_nt/schlaglichter_nt/article155779160/Viele-Verletzte-nach-Blitzeinschlaegen-in-Deutschland-und-Frankreich.html. Accessed on 25 July, 2017

Duclos S (2015) Is Texas under attack by the U.S. Government for resisting jade helm? "Weather warfare right before our very eyes". Retrieved from All News PipeLine on June 14, 2016: Retrieved on July 28, 2017, from http://allnewspipeline.com/Texas_Under_Attack_Resisting_Jade_Helm.php

Eastlund BJ (1987) United States patent 4,686,605. Method and apparatus for altering a region in the earth's atmosphere, ionosphere, and/or magnetosphere. In: [US Patent & Trademark Office, patent full text and image database]; august 11, 1987. Retrieved from http://patft.uspto.gov/netacgi/nph-Parser?Sect1=PTO1&Sect2=HITOFF&d=PALL&p=1&u=%2Fnetahtml%2FPTO%2Fsrchnum.htm&r=1&f=G&l=50&s1=4,686,605.PN.&OS=PN/4,686,605&RS=PN/4,686,605 on July 24, 2017

Eastwood C (2001) The cost of… …weather control. In: The Oberver online, August 12, 2001. Retrieved on December 15, 2017, from https://www.theguardian.com/theobserver/2001/aug/12/life1.lifemagazine

Ellsberg D (2017) http://www.biography.com/people/daniel-ellsberg-17176398. Daniel Ellsberg Biography.com. Scholar, anti-war activist, government official, journalist (1931–) on July 25, 2017

EuroNews (2016) Rising floodwaters in Germany and France claim more lives. Retrieved from EuroNews online on June 13, 2016: Retrieved from http://www.euronews.com/2016/06/02/rising-floodwaters-in-germany-and-france-claim-more-lives/ on July 28, 2017

Faber D (2015) Environment. Does the Paris agreement open the door to geoengineering? On: Berkely, December 14, 2015. http://blogs.berkeley.edu/2015/12/14/does-the-paris-agreement-open-the-door-to-geoengineering/

Fleming JR (2007) On the possibilities of climate control in 1962. Harry Wexler on geoengineering and ozone destruction. Science, technology, and society, Colby Collge, American geophysical union, December 14, 2007

Gathmann L (1891) United States patent 0462795. Inventor: Louis Gathmann, of Chicago, Illinois. Retrieved on June 5, 2017, from http://www.freepatentsonline.com/0462795.pdf

Hamblin JD (2013) Arming mother nature: the birth of catastrophic environmentalism. Oxford University Press, Oxford

Hap A (2013a) Project cirrus: 1940's weather modification. In: Opsec news, published on April 2, 2013. Retrieved from http://www.opsecnews.com/project-cirrus-1940%e2%80%b2s-weather-modification/ on July 21, 2017

Hap A (2013b) Operation Popeye: Weaponized Weather during Vietnam war. In: Opsec news, published on April 3, 2013. Retrieved from http://www.opsecnews.com/operation-popeye-weaponized-weather-during-vietnam-war/. On July 25, 2017

Hauser C (2016) 6 dead in Texas floods, and more rain is coming. Retrieved from New York Times online on June 14, 2016: http://www.nytimes.com/2016/05/31/us/6-dead-in-texas-floods-and-more-rain-is-coming.html?_r=0

House TJ, Near JB Jr, Shields WB (USA), Celentano RJ, Husband DM, Mercer AE, Pugh JE (1996) Weather as a Force Multiplier. Owning the weather in 2025. Department of Defense School (ed.)

IFI CLAIMS Patents Services (2017) Retrieved on July 15, 2017, from http://www.google.com/patents/US20030085296

IPCC (2014) Summary for policymakers. In: Field CB, Barros VR, Dokken DJ, Mach KJ, Mastrandrea MD, Bilir TE, Chatterjee KL, Ebi KL, Estrada YO, Genova RC, Girma B, Kissel ES, Levy AN, MacCracken S, Mastrandrea PR, White LL (eds) Climate change 2014. Impacts, adaptation, and vulnerability. Part A: global and sectoral aspects. Contribution of working group II to the fifth assessment report of the intergovernmental panel on climate change. Cambridge University Press, Cambridge, pp 1–32

Kieler Earth Institute (2017) Climate engineering news. Retrieved from http://www.climate-engineering.eu/news.html on July 22, 2017

Knabb RD, Rhome JR, Brown DP (2011) Tropical cyclone report. Hurricane Katrina. National Hurricane Center, Miami

Lambert KK (2011) Patent US 8033879 B2-Biophysical geo-engineering compositions and methods. https://www.google.com/patents/US8033879. Accessed on 15 July, 2017

Levitin C (1996) Russian documents set out 'tectonic weapon' research. In: nature 383, 471 (10 October 1996) | doi:https://doi.org/10.1038/383471a0. http://www.nature.com/nature/journal/v383/n6600/pdf/383471a0.pdf?foxtrotcallback=true. Accessed on 26 July, 2017

McElwee R (2016) Flash floods continue in France and Germany Slow-moving major thunderstorms cause flooding in central Europe where rivers overflow Retrieved from Al Jazeera online on June, 13, 2016: http://www.aljazeera.com/news/2016/06/flash-floods-france-germany-160602111600158.html

Musser WC (1960) US2963975 A. Cloud seeding carbon dioxide bullet. US2963975A http://www.google.com/patents/US2963975?hl=de&dq=2963975&ei=R21LULSTKYTY8gTL6YDgCA. Accessed on 25 July, 2017

NASA (2013). Record details. Retrieved from NASA Technical Reports Server-Providing Access to NASA's Technology, research, and Science online on June 14, 2016: http://www.ntrs.nasa.gov/search.jsp?R=19680002906

Newell HE (1966) A recommended national program in weather modification. A report to the interdepartmental Committee for Atmospheric Sciences. Retrieved from NASA Technical Reports Server-Providing Access to NASA's Technology, research, and Science online on June 14, 2016: http://www.ntrs.nasa.gov/search.jsp?R=19680002906; http://ntrs.nasa.gov/archive/nasa/casi.ntrs.nasa.gov/19680002906.pdf

NOAA (2015) New report finds human-caused climate change increased the severity of many extreme events in 2014. Retrieved from http://www.noaanews.noaa.gov/stories2015/110515-new-report-human-caused-climate-change-increased-the-severity-of-many-extreme-events-in-2014.html. On June 15, 2017

Odenwald M (2007) Kann man das Wetter manipulieren. In: Focus online; November 30, 2007. Retrieved from http://www.focus.de/wissen/weltraum/odenwalds_universum/odenwalds-universum_aid:227601.html

Ott K (2011) Die Kartierung der Argumente zur Klimamanipulation. Die Debatte um Geoengineering bzw. Climate Engineering (CE). In: Jahrbuch für Ökologie 2011, S. Hirzel Verlag Stuttgart

Planungsamt der Bundeswehr (2012). Future topic geoengineering. In: Planungsamt der Bundeswehr (ed.). Streitkräfte, Fähigkeiten und Technologien im 21. Jahrhundert. Retrieved from Planungsamt der Bundeswehr on June 13, 2016: http://www.planungsamt.bundeswehr.de/portal/poc/plgabw?uri=ci%3Abw.plgabw.service.studien&de.conet.contentintegrator.portlet.current.id=01DB010100000001%7C9D7ES5155DIBR

Pook S (2001) Deadly flood blamed on RAF rainmakers, telegraph, 31 August 2001. Retrieved on July 21, 2017, from http://www.telegraph.co.uk/news/uknews/1339046/Deadlyflood- blamed-on-RAF-rainmakers.html

Quinn M (2013) Death by lightning a danger in developing countries. Scientists and educators are confronting superstitions and poor infrastructure. In: National Geographic, November 4, 2013. Retrieved on July 12, 2017, from https://news.nationalgeographic.com/news/2013/11/131102-lightning-deaths-developing-countriesstorms/

Rice D (2017) It's been a stormy start to 2017 in the U.S., tying the 2nd-most natural disasters on record. In: USA today; July 17, 2017. Retrieved from https://www.usatoday.com/story/weather/2017/07/18/u-s-battered-2nd-most-natural-disasters-record-so-far-2017/485167001/ on July 26, 2017

Rich M (2011) Weather Warfare Introduction Retrieved from New World War online on July 23, 2017: http://www.newworldwar.org/weatherwar.htm

Rovella J (1995) Tropical cyclone disruption US 5441200A. Retrieved from: http://www.google.com/patents/US5441200?dq=5441200&ei=lE9LUIPzHIic9gST2oHYAw#v=onepage&q=5441200&f=false on July 12, 2017

Schwarz-Herion O (2005) Die Integration des Nachhaltigkeitsgedankens in die Unternehmenskultur und dessen Umsetzung in die betriebliche Praxis. Eine empirische Studie zur ökologischen und sozialen Verantwortung von Privatunternehmen. Shaker, Aachen

Schwarz-Herion O (2015) Urgent ecological problems for the NSP. In: Schwarz-Herion O, Omran A (eds) Strategies towards the new sustainability paradigm. Managing the great transition to sustainable global democracy. Springer International Publishing, Cham, pp 121–140

Shabecoff P (1988) Global warming has begun, expert tells senate. In: New York Times, June 24, 1988. http://www.nytimes.com/1988/06/24/us/global-warming-has-begun-expert-tells-senate.html. Accessed on 25 July, 2017

Shepherd J (2016) What does the paris agreement mean for geoengineering? Retrieved on July 12, 2017, from http://blogs.royalsociety.org/in-verba/2016/02/17/what-does-theparis-agreement-mean-for-geoengineering/

Siddle J (2001) US makes 'weather control powder'. In: BBC news, Thursday, 2 August, 2001, 01:45 GMT 02:45 UK. Retrieved on July 24, 2017, from http://news.bbc.co.uk/1/hi/sci/tech/1469610.stm

Smolker R, Ernsting A (2012) BECCS (Bioenergy with Carbon Capture and Storage): climate savior or dangerous hype? Retrieved on July 12, 2017, from http://www.biofuelwatch.org.uk/2012/beccs_report/

Spaleta S (2017) Major Winter storm seen from space – set to wallop East Coast. Retrieved on July 26, 2017, from https://www.livescience.com/58237-major-winter-storm-seenfrom-space-set-to-wallop-east-coast-video.html

Srinivas H (2015) The Indian Ocean tsunami and its environmental impacts. Retrieved on June 2, 2017, from https://www.gdrc.org/uem/disasters/disenvi/tsunami.html

Teller E, Wood L, Hyde R (1997) Global warming and ice ages: prospects for physics-based modulation of global change. 22nd international seminar on Planetary Emergencies, Erice (Sicily)

The Black Vault (2015) Project cirrus – Hurricane modification. In: Document Archive on March 26, 2015. Retrieved from http://www.theblackvault.com/documentarchive/project-cirrus-hurricane-modification/# on July 20, 2017

The Millennium Report (2015) Texas floods: a Geoengineered weather event using Chemtrails and HAARP. Retrieved from The Millennium Report online on June 13, 2016: http://themillenniumreport.com/2015/05/is-the-texas-rain-deluge-and-flood-a-geoengineered-event/

The Royal Society (2009) Geoengineering the climate. Science, governance, and uncertainty. Science Policy Centre, London

The Royal Society (2014) Climate change evidence & causes. February 27, 2014.The National Academies Press, Washington. https:doi.org/https://doi.org/10.17226/18730

Thornes JE, Pope FD (2014) Why do we need solutions to Global Warming? In: Hester RE and Harrison RM (eds) Issues in Environmental Science and Technology. vol. 38. Geoengineering of the Climate System. The Royal Society of Chemistry, Cambridge

Trademarkia (2003) DYN-O-GEL Trademark information, DYN-O-MAT, INC. Retrieved on December 15, 2017, from https://www.trademarkia.com/dynogel-76122115.html

UNOG (1976) Text of the convention. Convention on the prohibition of military or any hostile use of environmental modification techniques, 10 December 1976. http://www.unog.ch/unog/website/disarmament.nsf/(httpPages)/65BBB9DF7D12D5F1C1257FD9005C9C64?OpenDocument&unid=1C87969B8C841FCAC1257B11005C3E61. Accessed on 22 July, 2017

UNOG (1978) Convention on the prohibition of military or any other hostile use of environmental modification techniques (ENMOD). Retrieved from http://www.unog.ch/enmod. On July 22, 2017

US Agency for International Development (1998) Report Caribbean, Dominican Republic, Haiti – Hurricane Georges Fact Sheet #6. Retrieved from http://reliefweb.int/report/antigua-and-barbuda/caribbean-dominican-republic-haiti-hurricane-georges-fact-sheet-6 on July 21, 2017

USA Today (2001) People killed in plane attacks. In: USA Today online, 09/25/2001 – Updated 12:52 PM ET. http://usatoday30.usatoday.com/news/nation/2001/09/11/victims-list.htm. Accessed on 25 July, 2017

Uther S (2014) Diskurse des Climate Engineering. Argumente, Akteure und Koalitionen in Deutschland und Großbritannien. Springer VS, Wiesbaden

Vidal J, Weinstein H (2001) RAF rainmakers 'caused 1952 flood'. Unearthed documents suggest experiment triggered torrent that killed 35 in Devon disaster. In: The Guardian Online, August 30, 2001. Retrieved on July 14, 2017, from https://www.theguardian.com/uk/2001/aug/30/sillyseason.physicalsciences

Wall M (2017) Gigantic X-ray tsunami rolls through galaxy cluster (Video, Photos). On: space.com on May 3, 2017. Retrieved On July 25, 2017, from https://www.space.com/36698-x-ray-tsunami-perseus-galaxy-cluster-video.html

Waxmanski A (2003) Hurricane and tornado control device. US 20030085296 A1. Retrieved on July 15, 2017, from http://www.google.com/patents/US20030085296

Weart S (2017) The discovery of global warming - government: the view from Washington, DC. In: http://history.aip.org/climate/summary.htm. Retrieved from July 20, 2017

Weather Modification Inc (2017) When most people look up they see clouds. We see potential. http://www.weathermodification.com/. Accessed on 26 July, 2017

Westing AH (2013) Pioneer on the environmental impact of war. In: Brauch HG (ed) SpringerBriefs on pioneers in science and practice, vol 1. Springer, Heidelberg/New York/Dordrecht/London

Wexler H (1958) Modifying weather on a large scale. Science 128(3331), published on October 31, 1958

Williams R (2013) CIA backs $ 650, 000 study into how to control global weather through geoengineering. Study part-funded by the CIA to investigate national security investigations of geoengineering. http://www.independent.co.uk/news/world/americas/cia-backs-630000-study-into-how-to-control-global-weather-through-geoengineering-8724501.html. Accessed on 28 July, 2017

WMO (2017a) World highest mortality hailstorm. https://wmo.asu.edu/content/world-highest-mortality-hailstorm. Accessed on 3 June, 2017

WMO (2017b) WMO releases list of the deadliest weather events. Retrieved from https://www.earth.com/news/wmo-list-deadliest-weather-events/ on June 3, 2017

Index

A. Omran, O. Schwarz-Herion (eds.), *The Impact of Climate Change on Our Life*, https://doi.org/10.1007/978-981-10-7748-7

W

Zeitfracht Medien GmbH
Ferdinand-Jühlke-Straße 7
99095 Erfurt, Deutschland
produktsicherheit@kolibri360.de